ETHICS
FOR A SMALL PLANET

A Communications Handbook on
the Ethical and Theological Reasons
for Protecting Biodiversity

BIODIVERSITY PROJECT

Biodiversity Project
214 N. Henry Street
Suite 201
Madison, WI 53703
(608) 250-9876
Fax: (608) 257-3513
www.biodiversityproject.org

ISBN 0-615-12258-2.

♻ Printed on 100% post consumer,
elemental chlorine free recycled paper.

Madison, Wisconsin, November 2002
© Biodiversity Project

Table of Contents

About the Biodiversity Project 5

Acknowledgements 6

Preface: Why a Handbook on Communicating about Ethical and Theological Perspectives on Biodiversity?—*by Jane Elder* 8

Introduction: Unpacking the Ethical Toolbox—*by Jane Elder* 10

About the Authors 13

SECTION I
Why Should We Talk about Ethics, Values, and Biodiversity?

Why Should We Talk about Ethics, Values, and Biodiversity? —*by Jane Elder* 18

> Ethics in the Real World: Speaking in a Forgotten Language— *by Bob Perschel* 20
>
> Ethics in the Media: "Our View: Conservation is a Moral Cause" (from the *Idaho Falls Post Register*) 22

SECTION II
Origins and Roots: A Crash Course in Theological and Ethical Perspectives on Biodiversity

Biodiversity, Theology, and Ethics: Key Concepts—*by Peter W. Bakken* 24

> Ethics in the Media: Redwood Rabbis (from *SIERRA Magazine*) ... 27

Overview of World Religions —*by Jane Elder and Marian Farrior* ... 29

Biblical and Theological Perspectives on Biodiversity —*by Peter W. Bakken (with contributions from Daniel Swartz)* 32

> Ethics in the Media: Resolving Conflict in the Chesapeake Bay (from the Au Sable Institute Newsletter) 40

Introduction to Environmental Ethics—*by Michael P. Nelson* 41

2,000 Years of Western Ideas about Nature in Less Than 2,000 Words —*by Michael Nelson* 48

SECTION III
How Shall We Live? Applying Ethical and Religious Perspectives to the Biodiversity Crisis

The Ways We Value Nature —*by Michael P. Nelson* 56

Rights and Responsibilities: What Obligations Do We Owe to the Natural World (and Each Other)? —*by Michael P. Nelson and Robb Cowie* 61

> Ethics in the Real World: The Klamath Basin Controversy 62

Obligations to the Future —*by Daniel Swartz* 65

Murky Waters: When There's No Clear Line between the Right and Wrong Choices—*by Jane Elder* 68

✻ "In our every deliberation, we must consider the impact of our decisions on the next seven generations."

Gayanashagowa, *The Great Binding Law* of the Iroquois Confederacy

Right v. Right Conflicts: A Process
for Ethical Decision Making
—by Nancy J. Miaoulis 70

Ethics in the Real World:
The Maine Fisheries Dilemma 71

How Shall We Live?
Environmental Equity and Justice
—by Daniel Swartz 74

SECTION IV
Thinking Locally, Acting Globally: Steps Toward an Ethic for the Biosphere

Saving Land and People
—by Peter Forbes 78

Ethics in the Real World:
Difusing the Wise-Use
Movement 82

The Earth Charter: Guide to
a Sustainable Way of Life
—by Dieter T. Hessel 84

SECTION V
Communications Tips and Tools: Talking about Biodiversity, Ethics, and Faith

The Art of Communicating
about Ethics—by Jane Elder 92

Ethics in the Media:
Movement Connects the
Heavens with Earth (from
The Oregonian, Portland, OR) 98

Crafting and Using Values-Based
Messages—by Jane Elder 101

Busting Anti-Conservation Myths
—by Michael Nelson 104

Talking About Biodiversity
—by Jane Elder 107

Getting Ethical Messages Out
—by Jane Elder and Erin Oliver 109

Ethics in the Media:
Christians Tackle Climate
Change through International
Conference (from *The Capitol
Times*, Madison, WI) 112

Calendar of Events and
Opportunities 114

Ethics in the Media: Nature Needs
Saving on South Shore (from *The
Patriot Ledger*, Quincy, MA) 116

SECTION VI
Resources

Experts and Speakers on
Biodiversity and Ethics 120

Glossary of Terms 122

Bibliography 124

Additional Resources 129

Appendix I:
The Assisi Declarations 136

Appendix II:
The Earth Charter 140

> ✷ "The world is so beautiful, so utterly beautiful. Who else is turning with joy as the Earth is turning?"
> Terry Tempest Williams

About the Biodiversity Project

The Biodiversity Project is a nonprofit organization whose mission is to advocate for biodiversity by designing and implementing innovative communication strategies that build and motivate a broad constituency to protect biodiversity. We began as an initiative of the Consultative Group on Biological Diversity in 1995, and we became an independent nonprofit in 2000. We are based in Madison, Wisconsin, and work with organizations throughout the U.S. and Canada.

We are the only organization in the United States focused on building a long-term constituency for biodiversity and the many issues that affect it. To build broad and deep public participation in biodiversity conservation, we need to understand the values that underlie people's attitudes. Using public opinion and social science research, we strive to identify promising new constituencies for biodiversity conservation and to develop strategies for reaching those audiences in meaningful ways. We are catalysts and leaders, putting sophisticated communications tools into the hands of activists, educators, and public voices for biodiversity, helping them create effective messages with lasting impact.

* "Let ours be a time remembered for the awakening of a new reverence for life, the firm resolve to achieve sustainability, the quickening of the struggle for justice and peace, and the joyful celebration of life."

The Earth Charter

Biodiversity Project
Life. Nature. You. Make the connection.

Acknowledgements

This project was made possible through the support of The Henry R. Luce Foundation.

The Biodiversity Project thanks the many individuals and organizations that contributed to the contents of this kit:

Writers and Editors
Peter Bakken, Coordinator of Outreach and Research Fellow, Au Sable Institute
Robb Cowie, Managing Editor, Deputy Director, Biodiversity Project
Jane Elder, Executive Director, Biodiversity Project
Marian Farrior, Program Manager, Biodiversity Project
Peter Forbes, Vice President & National Fellow, Trust for Public Land
Charlotte Frascona, Copy Editor
Dieter Hessel, Director, Program on Ecology, Justice, and Faith
Nancy Miaoulis, Intern, Institute for Global Ethics
Michael Nelson, Associate Professor of Philosophy and Natural Resources, University of Wisconsin-Stevens Point
Erin Oliver, Production Editor, Communications Coordinator, Biodiversity Project
Barbara Sella, Copy Editor
Patricia Stocking, The Nature Conservancy
Daniel Swartz, Executive Director, Children's Environmental Health Network

Designer
Nancy Zucker, Zucker Design

Steering Committee Members
Peter Bakken, Coordinator of Outreach and Research Fellow, Au Sable Institute
Curt Meine, Wisconsin Academy of Sciences, Arts, and Letters
Kathleen Dean Moore, Professor of Philosophy, Oregon State University
Robert Perschel, Director, Land Ethic Program, The Wilderness Society
Carol Saunders, Director of Communications Research, Brookfield Zoo
Daniel Swartz, Executive Director, Children's Environmental Health Network

Advisors
Lisa Bardack, Coordinator, Earth Charter USA, Center for Respect of Life and Environment
Kathy Blaha, Senior Vice President for National Programs, Trust for Public Land
Dee Boersma, Professor of Zoology, University of Washington
Baird Callicott, Professor of Philosophy, University of North Texas
David Campbell, Professor of Biology and Henry R. Luce Professor of Nations and the Global Environment, Grinnell College
Cassandra Carmichael, Director of Faith-Based Outreach, Center for a New American Dream
Anne Custer, Research Associate, Forum on Religion and Ecology
Ron Engel, Research Professor of Environmental and Social Ethics, Meadville/Lombard Theological School, University of Chicago affiliate
Peter Forbes, Vice President & National Fellow, Trust for Public Land
Eric Freyfogle, Professor of Law, University of Illinois at Urbana-Champaign
Paul Gorman, Executive Director, National Religious Partnership for the Environment
Dale Jamieson, Professor of Philosophy and Henry R. Luce Professor of Human Dimensions of Global Change, Carleton College

> ✳ "A thing is right when it tends to preserve the integrity, stability, and beauty of the biotic community. It is wrong when it tends otherwise."
> *Aldo Leopold*

Stephen Kellert, Professor of Social Ecology, Yale University
Abby Kidder, Project Director, Institute for Global Ethics
Rhonda Kranz, Program Manager, Ecological Society of America
Suellen Lowry, Outreach Coordinator, Earth Justice Legal Defense Fund, and California Director, Interfaith Partnership for Children's Health and the Environment
Gary Meffe, Professor of Wildlife Ecology and Conservation, University of Florida, and Editor, *Conservation Biology*
Tom Muir, U.S. Geological Survey
Gene Myers, Professor of Geography and Environmental Social Sciences, Western Washington University
Michael P. Nelson, Associate Professor of Philosophy and Natural Resources, University of Wisconsin-Stevens Point
Bryan Norton, Professor of Philosophy, Georgia Institute of Technology
Max Oelschlaeger, Professor of Philosophy, University of North Texas
Edwin Pister, Executive Secretary, Desert Fishes Council
Holmes Rolston, Professor of Philosophy, Colorado State University
Michael Schuler, Pastor, First Unitarian Society, Madison, Wisconsin
Bron Taylor, Professor of Religion and Environmental Studies, University of Wisconsin-Oshkosh, and Editor, *Encyclopedia of Religion and Nature*

Additional Advisors
Meg Domroese, American Museum of Natural History
Susan Dowds, New England Aquarium
Ed Lytwak, Endangered Species Coalition
Sarah Matsumoto, Endangered Species Coalition
Mary Paden, GreenCOM and North American Association for Environmental Education
Debra Shore, Chicago Wilderness

Interns and Volunteers
Neda Arabshahi
Beverly Fowler, O.P.

Reprint Permissions
Bizarro cartoon, ©Dan Piraro Reprinted with special permission of Universal Press Syndicate
Christians Tackle Climate Change through International Conference, used by permission of Gordon Govier
I Need Help cartoon, © Vic Lee. Reprinted with special Permission of King Features Syndicate
Major Religions of the World Ranked by Number of Adherents, used by permission of Adherents.com
Movement Connects the Heavens with Earth, used by permission of Todd Wilkinson
Mutts cartoon, © Patrick McDonnell. Reprinted with special Permission of King Features Syndicate.
Nature Needs Saving on South Shore, used by permission of *The Patriot Ledger,* Quincy, MA
Our View: Conservation is a Moral Cause, used by permission of the Idaho Falls *Post Register*
People and the Land, used by permission of Peter Forbes
Redwood Rabbis, used by permission of Seth Zuckerman
Resolving Conflict, used by permission of Susan Drake Emmerich
The Assisi Declarations, used by permission of the United Nations Environmental Program
The Earth Charter, used by permission of the Earth Charter Initiative

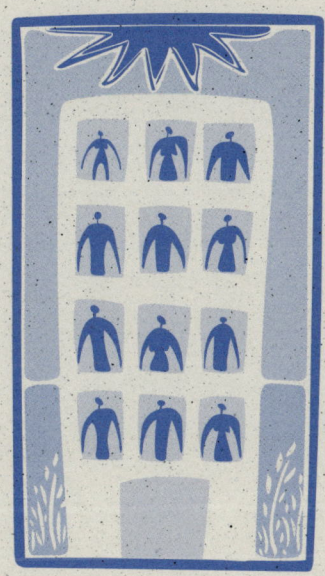

Preface: Why a Handbook on Communicating about Ethical and Theological Perspectives on Biodiversity?

by Jane Elder

Today, we're losing species and habitat at rates unparalleled in human history. What's more, Earth's life support systems—those that provide our clean air, fresh water, those that pollinate our food crops, those that regulate our climate, those that provide us with wonder, beauty, inspiration—are being degraded. Unless more people grow to understand what is at risk—unless more of us take action to sustain and protect life—the picture for humans and our fellow passengers on this small planet won't be very pretty.

At the Biodiversity Project, we believe that long-term solutions to the biodiversity crisis will come from an ethical conviction that saving biodiversity is the right thing to do. Science will help inform the discussion, yes. But when it comes to making tough decisions about how six billion of us will choose to live on Earth, we will need to dig deeply into values—our collective sense of right and wrong concerning each other, the generations that follow us, and the life with which we share this planet. Advocates for biodiversity will need to use all the tools in the social change toolbox if we hope to succeed in protecting biodiversity. One of these important tools will be the skill to communicate effectively about the moral dimensions of this issue.

What This Handbook Is, and What It Isn't

This handbook is meant to be a tool to help biodiversity spokespersons—activists, scientists, educators, and anyone else who loves the living planet—open a broader conversation with the public on the ethical issues related to protecting species, habitat, ecosystems, and all the interconnections that make our planet life-giving, wondrous, and beautiful. Our goals in creating this handbook are to help biodiversity advocates:
- Understand the origins of contemporary views on biodiversity conservation;
- Integrate ethical messages into their outreach; and
- Feel confident and credible doing so.

At the same time, this handbook is not a comprehensive analysis of all the ethical issues in the realm of the multi-faceted biodiversity crisis, nor was that our intent. New issues surface every day, and ethics are almost certainly at play in every issue that touches on the biodiversity crisis. However, we do explore many of the key ethical issues within this handbook, because real examples make abstract concepts easier to grasp and apply.

The handbook provides summaries and background on many aspects of environmental ethics and theological viewpoints related to the environment and biodiversity. It is an overview. There is a rich body of scholarly work on this topic, and we invite those interested to explore those original sources further.

We ask latitude from the many experts whom we consulted on this project, knowing that these complex topics can't be easily reduced to simple shorthand summaries. We have made an effort to respect the integrity of the subject we have tackled, without forgetting that our primary focus is on communicating about biodiversity and ethics, not providing another scholarly treatment of environmental ethics itself.

Any time a writer attempts to summarize something as complex and subjective as a worldview or a religious perspective in a brief essay, there is risk. We have attempted to treat a wide range of perspectives with respect and sensitivity, and ask the reader to consider that our aim was to illustrate, not to define these perspectives. We have

---

* This handbook is meant to be a tool to help biodiversity spokespersons—activists, scientists, educators and anyone else who loves the living planet—open a broader conversation with the public...

also made an effort not to critique any particular tradition or viewpoint as "good" or "bad" for biodiversity protection. First, it isn't the purpose of this project, and second, we expect that readers will draw their own conclusions based on their own values.

This handbook is an invitation to explore and test the waters of public dialogue about what is right and wrong for protecting biodiversity. Psychologists tell us that it is very difficult to change peoples' values once they are established. But we can "activate" values by helping make the connection between peoples' values and an issue they care about. In order to make that connection and provide a catalyst for action, it helps to understand the context and history of values and ethics in our culture. It also helps to understand what motivates people, how to talk with them about these topics, how the media look at covering these issues, and where the opportunities exist to enter into this kind of conversation.

We hope this resource helps you become a more effective voice for biodiversity, and we hope it encourages you to express the ethical principles that motivate you to protect and care for life on Earth. We invite you to explore these topics, sort through your own ethical framework, and consider whether ancient Greek philosophy, or scripture, or Deep Ecology has shaped the way you see nature today. We hope this kit helps you better understand the ethical "battle ground" over biodiversity, and also helps you to reach more people about the "why" of saving biodiversity, as well as the "what."

A Note of Thanks

We owe an enormous debt to our steering committee—each of them thinkers and leaders in their fields—who contributed to the conceptual design and content of this handbook.

We are also deeply grateful to the Henry R. Luce Foundation for this rare opportunity to explore the links between biodiversity, theology, and ethics and to help others become confident speaking about their values and their beliefs about how to act toward the rest of nature.

The publication of this handbook on 100 per cent post consumer recycled content paper was made possible by a generous contribution from the Fox River Paper Company. Not only do we consider such paper to be representative of our mission to protect biodiversity, we make the decision to publish on this paper because it is simply the right thing to do.

* We hope this resource helps you be a more successful, more effective voice for biodiversity, and we hope it encourages you to express the ethical principles that motivate you to protect and care for life on Earth.

Introduction: Unpacking the Ethical Toolbox

> *Our goal is to help you gain confidence and skill in expanding the public debate about biodiversity to include the ethical dimension. Why? Because we think it will help us all be more successful in protecting and saving biodiversity.*

This handbook is designed to help people who communicate about biodiversity to the public, whether it is through a community planning process, a policy debate, the media, or among colleagues and coalitions. Our goal is to help you gain confidence and skill in expanding the public debate about biodiversity to include the ethical dimension. Why? Because we think it will help us all be more successful in protecting and saving biodiversity.

We've dedicated a large share of this handbook to background, because most of us won't take on a complex issue in a public setting unless we feel like we know what we're talking about, and the ethical dimension of biodiversity is a topic with a lot of weight and nuance. In order to enter into a conversation that addresses values, ethics, and beliefs, you want to know the turf on which you are treading, and in this case that turf is comprised of thousands of years of human thought, tradition, debate, and exploration.

We've tried to provide a quick overview of context, history, and cultural interpretations of philosophy and theology that relate to biodiversity. At the same time, we felt it was important to free this topic from the ivory tower and cloistered halls and encourage others to infuse ethical perspectives into their outreach on biodiversity. Each of us acts through a personal framework of values and ethics every day, and it is through that framework that countless decisions about biodiversity are made. The more we understand ethical frameworks, the easier it will be to identify those occasions when it is important to shed light on things that are often relegated to instinct, or just the way things are. Maybe our ethical responses are instinctive, but where do those instincts come from? Maybe a situation appears to be just the way things are, but is it—can we frame it in a new light? So, we've combined the background from scholars and religious thinkers with practical tips on how to raise this topic in public communications. We hope this makes the ethical dimension of biodiversity issues a bit more transparent, and also provides you with some tools on how to bring it into play to connect with people and to craft solutions for tough issues.

Each section tackles this challenge from a different angle.

Section I: Why Should We Talk about Ethics, Values, and Biodiversity?

This section explores the reasons why talking about biodiversity in the context of ethics, values, and moral perspectives has a positive and powerful impact, and why today's issues require us to rise to a new level of ethical debate. It also includes examples of how raising the conversation to this level changes the tone of the debate and fosters solutions. The sound bite? The solution to today's biodiversity crisis requires human culture to rise to an unprecedented challenge, and much of that challenge involves values and ethical choices.

Section II: Origins and Roots.

Every good communicator does background research. Section II is a first step—what are ethics, what is theology, how are they connected—along with a brief overview of the diverse religious perspectives that relate to biodiversity (primarily those that are dominant in American culture). It also will help you understand the language, history, and main schools of thought in environmental ethics by providing an introduction to the field and its perspectives on biodiversity.

Section III: How Shall We Live?

This section explores how ethics and religious beliefs are currently applied to biodiversity. The world's major philosophical and religious schools of thought have offered us hope, wisdom, and inspiration to stem the biodiversity crisis. They also lie at the root of our environmental problems. Ethicist Michael Nelson and Rabbi Daniel Swartz offer different, but complementary observations on the ways we value nature and the compacts and covenants that govern our relationship to the land and each other. Jane Elder explores the murky place where issues don't fall neatly into "good and bad" choices. Nancy Miaoulis examines the model proposed by the Institute for Global Ethics (IGE) for wrestling with right vs. right issues and, in an accompanying interview, IGE's Abby Kidder talks about how to communicate with the public, policy-makers, and even the opposition in such difficult situations.

Our hope is that this section provides you with some "ah-ha!" moments, where you can recognize the ethical influences in the issues you are dealing with—deconstruct them if you will—and gain new insights into why many in our culture have made some of the choices they have. We also hope that it provides you with insights into how you can connect more effectively with people by better understanding what motivates them.

Section IV: Thinking Locally, Acting Globally.

Yes, that's a twist on the original "think globally, act locally" message, because these days we have to think and act on both levels and in between. In this section, Peter Forbes looks at rebuilding our human relationship with the land and each other and offers hopeful stories of communities who are doing some creative and effective local thinking that is leading to positive change. Bob Perschel recounts how sharing his personal story and his land ethic helped defuse a confrontation with an angry group of property rights activists and built relationships that lay a foundation for larger land conservation victories in the Northern Forest.

At the global scale, the Earth Charter represents a remarkable achievement—a multi-cultural, world-embracing statement of ethical living to sustain humans and the environment. Dieter Hessel provides a tour of this important charter (see Appendix II for the Charter itself).

Each of the essays in this section is here to answer the question, "is anybody actually doing anything with this ethics stuff, and is it having an impact?"

Section V: Communication Tips and Tools.

This section offers practical advice on how to communicate about ethics, values, and biodiversity. Jane Elder offers tips and reminders on how to integrate the power of ethical arguments into your outreach in a sensitive way, as well as a primer on using values-based communications. Michael Nelson provides a handy "point/counterpoint" piece to help you combat some of the myths and tired old arguments against biodiversity protection. And we've put together recommendations on getting your ethical message out to the public through the media and other avenues. This section includes a calendar of holidays and special occasions that provide opportunities for holding events where you can talk about ethics and biodiversity protection.

Section VI: Resources.

The first thing you'll encounter in this section is a list of experts you can call on who can talk about why saving biodiversity is the right thing to do. We've also included handy, short glossary. It will help you sort out many of the terms in this Handbook. If you want to explore the topic further, this section is filled with additional resources, contacts, and Web sites. It includes a bibliography of sources used in compiling this Handbook.

Finally, throughout this Handbook you will find news clips and case studies on the way the ethics-biodiversity connection is being played out in the media and suggestions for getting further coverage on the topic.

About the Authors

Peter Bakken
Peter Bakken is Coordinator of Outreach and a Research Fellow for the Au Sable Institute in Madison, Wisconsin. He earned his B.A. in English and Religion from Concordia College, and his Ph.D. in Theology from the University of Chicago. His dissertation was titled "The Ecology of Grace: Ultimacy and Environmental Ethics in Aldo Leopold and Joseph Sittler." He was a member of the task force that produced the Evangelical Lutheran Church in America social statement, "Caring for Creation: Vision, Hope and Justice." His publications include *Ecology, Justice and Christian Faith: A Critical Guide to the Literature* (co-compiled with J. Ronald Engel and Joan Gibb Engel; Greenwood Press, 1995) and *Evocations of Grace: Writings on Ecology, Theology and Ethics by Joseph Sittler* (co-edited with Steven Bouma-Prediger, Eerdmans, 2000). He was born and raised in Fargo, North Dakota, and currently lives in Madison, Wisconsin, with his wife and daughter.

Jane Elder
Jane Elder is Executive Director of the Biodiversity Project. She brings more than twenty years of experience in the environmental movement to the Project, with expertise in both policy and communications. Before coming to the Project, Jane served as Director of Ecoregion Planning for the Sierra Club, heading a program to develop strategies for restoring and protecting the major regional ecosystems of the United States and Canada. She also headed the Sierra Club's Midwest regional office and its Great Lakes Program for many years. While at the Club, she played a central role in developing policy to regulate toxic air pollution, nationwide, along with other measures to protect the Great Lakes ecosystem. She also played a central role in campaigns to establish federally protected wilderness areas in the Midwest. She earned a Master's in Land Resources (1991) from the University of Wisconsin, Institute for Environmental Studies. Her research explored the role television production practices play in framing environmental issues. She also has a B.A. in Communication Arts (1976) from Michigan State University. She and her husband Bill Davis have a son, Colin Davis. They live in Madison, Wisconsin.

Peter Forbes
Peter Forbes is a writer, photographer, conservationist, and farmer. He is also Director of the Center for Land and People. For ten years, Peter oversaw the land conservation work undertaken by the Trust for Public Land (TPL) in New England. His major achievements at the Trust include working with the Walden Woods Project to protect threatened portions of Thoreau's Walden Woods; protecting and revitalizing urban gardens and farms in Boston, Massachusetts, and Providence, Rhode Island; adding 10,000 acres of wildlands to New Hampshire's White Mountain National Forest; and

creating the Good Life Center in Harborside, Maine, to promote the life and work of renowned authors and social activists Helen and Scott Nearing. In 1998, Peter became the Trust for Public Land's first National Fellow. His work is rooted in his belief in the radical center of a new expression of the human spirit in the natural world. Peter's photographic images and written words have appeared in more than a dozen books. He is the editor of *Our Land, Ourselves: Readings on People and Place* (Trust for Public Land, 1999) and, most recently, the author of *The Great Remembering* (Trust for Public Land, 2001). Peter serves on the Board of Directors of the Center for a New American Dream and the Vallecitos Mountain Refuge. He lives and farms in the Mad River Valley of Vermont.

Dieter Hessel

Dieter Hessel is a Presbyterian theologian and minister, specializing in social ethics and education. Since 1993, he has been Director of the ecumenical Program on Ecology, Justice, & Faith in Princeton, New Jersey, a professional development program for scholars and teachers in theology and religion. He is also Co-director of Theological Education to Meet the Environmental Challenge, a joint project with the Center for Respect of Life and Environment, Washington, D.C. He completed his Ph.D. studies in Christian social ethics at the Graduate Theological Union and San Francisco Theological Seminary in 1966, and received a B.A. degree (cum laude) from the University of Redlands and a B.D. from San Francisco Theological Seminary. Dieter is the founding co-chair of the Eco-Justice Working Group, National Council of Churches, and directed the Presbyterian Church USA unit that produced Restoring Creation for Ecology & Justice (1990). He is the author of several books and numerous articles on peace, social justice, and social ministry, and is the editor of many anthologies, including *Earth Habitat: Eco-Injustice and the Church's Response* (co-edited with Larry Rasmussen, 2001), *Christianity and Ecology: Seeking the Well-being of Earth and Humans* (co-edited with Rosemary Ruether, Harvard Center for the Study of World Religions, 2000), and *Theology for Earth Community: A Field Guide* (Orbis Books, 1996). He and his wife, Karen McLean Hessel, have three children and two grandchildren.

Nancy J. Miaoulis

Nancy J. Miaoulis, a certified trainer with the Institute for Global Ethics, is a graduate student at the University of New Hampshire. She is pursuing her doctoral degree in Environmental Ethics. An ordained minister in the United Church of Christ, she is currently on staff as a minister at The Second Congregational Church in Newcastle, Maine. She lives in Newcastle, Maine with her husband George, son Nicholas, and their yellow lab Sunny.

Michael P. Nelson

Michael P. Nelson is Associate Professor of Philosophy and of Natural Resources (joint appointment) at the University of Wisconsin-Stevens Point. Michael holds a B.A. from the University of Wisconsin-Stevens Point, an M.A. from Michigan State University, and a Ph.D. from Lancaster University, England, all in philosophy. Michael has published a number of articles in the area of environmental ethics and is co-editor, with J. Baird Callicott, of *The Great New Wilderness Debate*

(University of Georgia Press, 1998). He is currently working on a collection of the wilderness papers of Aldo Leopold and essays of contemporary commentary on Leopold's wilderness legacy. He is also writing a book tentatively entitled Ojibwa Land Ethics with Baird Callicott for Prentice-Hall.

Bob Perschel

Bob Perschel is Director of the Wilderness Society's Land Ethic Program. Before taking over this position in 1999, Bob served as Northeast Regional Director in the organization's Boston office. He received his B.A. in Psychology from Yale College and a Master's degree in Forestry Science from the Yale School of Forestry and Environmental Studies. He spent 15 years in the field as a forester for the forest industry and as a principal in his own consulting business. He was the founder of the Quinebaug Rivers Association, a Connecticut-based river protection organization. Bob is also co-founder of the Land Ethic Institute and was involved in the Society of American Forester's adoption of a land ethic in its code of ethics. Bob served as chair of the Northern Forest Alliance, a coalition working to protect the Northern Forest. He is also co-leader of The Wilderness Society's Land Ethic Task Force, which is responsible for developing programs to foster the connection between people and place.

Daniel Swartz

Rabbi Daniel Swartz serves as Executive Director of the Children's Environmental Health Network. Formerly, he was the associate director of the National Religious Partnership for the Environment (NRPE). As a part of his work with NRPE, Daniel oversaw its Washington, D.C., office and its public policy work and helped to address key environmental debates from a religious perspective. He lectures widely on religious perspectives and environmental issues, and is the author of numerous materials used by Jewish social activists working for justice. Before his work with NRPE, Daniel served as Congregational Relations Director for the Religious Action Center of Reform Judaism, the Washington office and social action arm of the Reform Jewish Movement, which is the largest Jewish organization in North America. Prior to coming to Washington, Daniel served for three years as Assistant Rabbi of Temple Israel of Hollywood in Los Angeles, where he directed one of the largest programs for welcoming Jews from the former Soviet Union. While in Los Angeles, Daniel helped found the L.A. Jewish Feminist Center and L.A. Works, a volunteer clearinghouse. Daniel was ordained from the Hebrew Union College-Jewish Institute of Religion in 1990, after receiving a Master's in Hebrew Letters from HUC-JIR in 1988. Daniel earned his B.A. (magna cum laude) in Environmental Studies from Brown University, where he was awarded the Senior Prize in Environmental Studies. He currently resides in Maryland with his wife and daughter.

Contributing authors **Robb Cowie, Erin Oliver** and **Marian Farrior** are Biodiversity Project staff members.

SECTION I

Why Should We Talk about Ethics, Values, and Biodiversity?

Why Should We Talk about Ethics, Values, and Biodiversity?

by Jane Elder

Introduction

Why step into the murky world of values, ethics, moral perspectives, and theological viewpoints? Why not stick to the facts, the purely rational? Why? Because humans are complex beings, and we make decisions about what to do, about what is right and wrong, through a mix of thought and feeling, rational argument and intuition, head and heart, data and gut instinct.

Whenever environmental problems are debated, there is almost always someone who encourages environmental advocates to set aside personal feeling and focus on the facts. Facts are useful, but they aren't the whole of the matter. All the data in the world won't persuade you to do something if some inner voice is screaming, "That's wrong!" Underlying that inner voice—the one that tells you, "this is right, this is important, trust this, but not that," your sense of right and wrong—is a tapestry of ethical history, cultural norms, and personal values. This communications kit is about exploring that tapestry and better understanding what makes people (Americans in particular), come to conclusions about what they value in relation to biodiversity, how those values shape decisions about biodiversity, and how we can better address those values and weave ethics into our outreach.

Four Reasons to Talk about Ethics. Here are four reasons why understanding and addressing the moral and spiritual basis of biodiversity protection can make us more effective in expanding the dialogue about biodiversity and fostering solutions to current problems.

1. **Most lasting social change is anchored in a deep moral imperative.** Throughout human history, human society has demonstrated its capacity for positive change, whether it is rejecting slavery, increasing global awareness of basic human rights, or enacting child labor laws. While social change is played out on a complex stage of economic, social, and political dynamics, at the heart of progress in human culture is a driving force that the change is the right thing to do—the moral imperative. Increasingly, leading thinkers in the field of biodiversity—from E.O. Wilson to Jane Goodall to David Suzuki—have given voice to the moral dimensions of the biodiversity crisis. Biodiversity advocates must expand the dialogue to explore and claim the

* Scratch just about any biodiversity issue, and under the surface is a collection of ethical issues waiting to be addressed.

moral imperative to protect Earth's life support systems. What larger moral question have we faced, if not the future of our species and the rest of life on Earth? If we fail to tackle the moral and ethical aspects of the biodiversity crisis, the "Sixth Great Extinction" (the current and largest loss of species in human history) will remain primarily the concern of science and academia. People will view it as a problem for experts to solve, not all of us.

2. Values-based rationales for protecting biodiversity are widely held and persuasive. Protecting biodiversity is important to people because they believe that they have a responsibility to protect the Earth for future generations, and because they believe that nature is God's creation and that they should respect the work of God. In the Biodiversity Project's 2002 and 1996 polls, these two reasons outranked all others in the survey.

In our 2002 poll, a majority of those surveyed strongly agreed with one of these two statements in a split sample question: "we have a *moral* responsibility to protect biodiversity" (65%) or "we have a *personal* responsibility to protect biodiversity" (69%).

Values are the lenses through which we judge the information we receive. Factual data—including scientific and economic statistics—are interpreted, shaped, and evaluated through the prism of our values. Therefore, values are as important, if not more so, than facts in getting a message across.

3. Reframing the debate humanizes and personalizes choices about biodiversity. If protecting biodiversity is about protecting my quality of life and my child's future (or yours and your child's), then it isn't about Latin names of creepy critters that only someone with a Ph.D. could love—it's about me and you and the people we care about. Suddenly, it's a personal issue. The more we can humanize biodiversity, and the less it remains in the realm of numbers, science, and something that happens "out there in nature," the better chance we have of connecting with people of diverse backgrounds and interests. Addressing the ethical dimensions of a biodiversity debate—what's right for my family and the generations that follow us?—changes the conversation.

4. Understanding ethics will help us make better decisions on complex issues. Many of the questions we face about biodiversity aren't going to come with easy answers as to which choice is the best, the most effective, the most likely to succeed. The more we understand about the ethical roots in our diverse and increasingly global culture, the more easily we can assess the complex issues that arise out of the context of our history and our present environmental crisis. Charting a pathway for biodiversity protection through multiple value systems, political systems, and the best of what science can tell us isn't easy. By increasing our own awareness of the value systems at play, we can better understand the playing field upon which biodiversity decisions will be made.

A sustainable future for humans and for the living Earth will need to address questions of equity, justice, compassion, and choices between near-term human needs and long-term human and biological needs. These ethical issues and questions are being forced by the scale, complexity, and urgency of the problems we face related to biodiversity. Here is a quick look at just a few of the key issues where our ability to communicate about ethical challenges will be crucial to protecting biodiversity.

Issues of a Global Scale. Issues such as climate change, the ozone hole, migratory species protection, deforestation, coral reef destruction, etc., all require cooperation across nations, economic systems, and cultures. If we fail to work together now, then we condemn all parties to grappling individually with these environmental problems as they get worse, and biodiversity and people will almost certainly suffer.

When asked to choose the single most important reason that they favor environmental protection, Americans identify the value of responsibility to future generations in greatest numbers. Nearly four in ten (39%) cite this as their main motivation, followed by respect for nature as God's work (23%). The desire to protect the balance of nature ranks third at 17%.

Choosing the Most Important Reason to Protect Environment

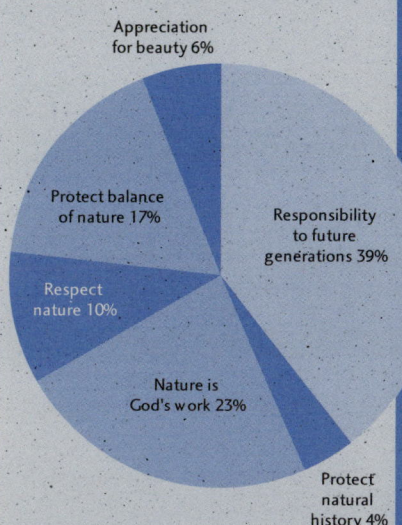

Q34. If you could choose only one of these six reasons, which one would you say is most important to you personally as a reason for you to care about protecting the environment: a) nature is God's work, b) respect nature for its own sake, c) to protect the balance of nature for you and your family to enjoy health, d) appreciation for nature's beauty, e) responsibility to future generations to protect the earth, and f) to protect America's natural history.

Source: Beldon, Russonello and Stewart, *Americans and Biodiversity: New Perspectives in 2002* (The Biodiversity Project: Madison, WI, 2002), 17.

WHY SHOULD WE TALK ABOUT ETHICS, VALUES, AND BIODIVERSITY?

* *Addressing the ethical dimensions of a biodiversity debate—what's right for my family and the generations that follow us?—changes the conversation.*

Solutions to these global challenges will come in part from common agreements that must be forged from diverse views of the world and divergent cultural values. If we can better understand the diversity of ethical views, and gain confidence in how to talk about them in the context of policy, we can play a valuable role in moving solutions forward.

Globalization and Industrial-scale "Harvesting" of Natural Resources. The facelessness of key players in the biodiversity crisis is another hurdle to overcome in shaping ethical solutions. In a village or a neighborhood, the community reinforces behaviors that benefit the community. However, globalization and industrial-scale production take the neighbor out of the equation. Most Americans see only the end product of globally produced goods (and perhaps the packaging it comes in) and know little about the process or the impact of production. If the environmental impacts of production are invisible to us, it is hard to identify and respond to the ethical issues or the biological or social ones. A challenge and opportunity for biodiversity advocates is to make the invisible visible so that an ethical response is possible. When we can name the threat, the actors, and the ethical issues at play and propose a responsive action, we can move from overwhelmed to active.

In a similar light, under U.S. law, corporations are treated as individuals, but as we all know, a corporation is not an individual, and we cannot interact with one in the way we would with a neighbor or another person. Corporations also don't make decisions

---

ETHICS IN THE REAL WORLD

Speaking in a Forgotten Language: Communicating Personal Ethics to Save Forests

by Bob Perschel, The Wilderness Society

The Northern Forest Lands Council (NFLC) was established by Congress in 1990 to develop recommendations on the future of the Northern Forest, a 26 million acre swath of mountains, forests, and small, struggling communities stretching from northern Maine to New York's Adirondack Park. In the early 1990's, the Council held a series of public meetings to gather comments for its final report. These sessions often became battlegrounds between environmentalists, property rights activists, and others.

In 1994, after an important Congressional hearing in the region was dominated by an aggressive turnout of Wise Use activists, some of us at The Wilderness Society (TWS) decided to try speaking out in a different voice:

Goals:
1) Send a clear message to the NFLC and Congress that they needed to improve the plan to protect the Northern Forest.
2) Dampen the influence that Wise Use rhetoric was having on the NFLC planning process.

Strategies:
We thought we could change the tone and nature of the debate—and blunt the thrust of Wise Use propaganda—by encouraging wilderness advocates to stand up and deliver their own "land ethic" to the Council and the media. The Wilderness Society was in the middle of a major economic study of the Northern Forest. But we wanted to inspire our troops and the public to speak to what was most meaningful to us, to articulate what brought us to the Northern Forest in the first place.

To accomplish this, first we trained activists to speak about their personal connections to the Northern Forest and the ethical underpinnings of their desire to protect it.

Second, we acted on the belief that if you want change, you have to represent change. TWS staff modeled this "land ethic" approach by speaking about our own values and attachments to the land, and we wove the land ethic into our public testimonies, presentations, and media work. We also placed an article in the Northern Forest Forum entitled, "Forgotten Language: Giving Voice to the Northern Forest," that talked about the challenge before us. In the article, I wrote:

"How is it that our arguments for the quality of human life seem always to be joined with economy . . . with that which is so completely utilitarian? How is it that we allow the Northern Forest debate to continue without summoning ourselves and others to that which is most meaningful to us?

We are asking the public to see the

through the ethical frameworks and grounding that most people carry around in their heads. Instead, many of the choices corporations make are primarily based on financial performance. Finding the individuals who are responsible for a corporation's ethical (or unethical) actions and naming names puts a face on that which is largely faceless and enables us to address corporate environmental ethics human-to-human, neighbor-to-neighbor on this small and crowded planet.

Privatization of Nature. To a large extent, the privatization of nature is threatening the concept of the commons, of basic human rights, and of access to basic and central ecological needs. The notion that living ecosystems or vital portions thereof can be parceled out as units of "ownership" (especially when there are no obligations to that which is owned or to others who are affected by that ownership) presents an ethical dilemma on a global scale. Likewise, applying notions of sovereignty and private property to biosphere-scale ecosystem services, such as air quality and fresh water (or individual life forms and genetic information), presents thorny environmental and social problems of equity, justice, and the protection of and access to the common good that biodiversity provides for us all.

This is just the tip of the "ethical iceberg" in the arena of issues waiting to be addressed. We need to be able to identify and talk about these issues because it will help is find solutions.

Northern Forest in a new way, to make decisions about its future from a new perspective. If we wish this to happen, we must first allow the change to occur within us. It is not enough for us to engage in the set parameters of the current debate. We must lead by going beyond them to introduce a more expansive way of dealing with our relationship to this forest—we must begin to speak in a forgotten language, one that has the power to convey our science and economics as well as our emotions, intuitions, and intimate connections to the land."[1]

Third, in our preparation for the next meetings, we made it a TWS priority that all speakers deliver the most personal message about their love for the land that they were capable of articulating. In our written instructions, we urged them: "Tell your story. We will use scientific and economic data, but it won't be enough. So at the beginning of your statement spend a few seconds speaking from the heart about the Northern Forest." We also invited our allies to do the same.

Results:
3,000 people attended the next public meetings, 700 spoke, and 77% spoke out in favor of our conservation agenda. Many of them did what we asked and presented their land ethic through personal stories of their connection to the forest. The press picked this up and quoted from these stories in their coverage of the meeting. Here's an example—a lead from an article in the *Christian Science Monitor*:

"It's a spiritual place. I can smell the mysteries of the moss."

"I remember rafting down the Penobscot—gurgling near the riverbanks—seeing osprey [and] moose."

"What we need to do: buy massive amounts of land for ecosystems . . . You can fly over Maine and not leave a clear cut in half an hour. The game is over. The forest is mostly gone."

These are the picturesque but sometimes urgent words of . . . New England residents who voiced their views on the Northern Forest . . .[2]

What the press and the public witnessed at these sessions was a genuine outpouring of emotion for their forests, as well as forceful demands to protect them, based on the personal ethics of the hundreds of people who testified. The Wise Use movement was neutralized by a richer, deeper, more encompassing set of values and passions.

Notes
[1] Robert Perschel, "Forgotten Language: Giving Voice to the Northern Forest," *Northern Forest Forum* (date unknown).

[2] Shelby Siems, "Protecting the Future of the Northern Forest," *Christian Science Monitor*, May, 1994.

ETHICS IN THE MEDIA

Our View: Conservation is a Moral Cause

by J. Robb Brady
Idaho Falls *Post Register*
March 7, 2002

Protecting the environment is not just a political cause. It's a religious one as well.

This commitment to the environment is recognized by Mormons, Catholics, Protestants, Jews, Indian tribes and even the eastern world's holiest, the Dalai Lama.

"The environment is a moral issue," says Brigham Young University professor JoAnn Myer Valenti, a scholar of natural resources and mass communications. "And that is what needs to be communicated."

BYU ethnobotanist Paul Cox adds, "Conservation is ultimately a spiritual issue."

Elsewhere, several Protestant churches have taken up the duty to defend the environment. They have sponsored national forums and published position papers on environmental issues.

Last year Roman Catholic bishops from Idaho, Montana, Oregon, and Washington issued an "International Pastoral Letter" underscoring the spiritual importance of preserving the special environmental values of the entire Columbia Basin watershed. The report was the culmination of a penetrating study of the basin. Rangeland and forest health, water quality, special riparian and wildlife needs are all targeted in the bishops' letter. Responsibility for environmental stewardship was stressed throughout the report.

BYU's Valenti's findings on environmental attitudes deserve attention in Idaho. They include:

• Most Americans are pro-environment, as are most Idahoans, in survey after survey.

• Evidence shows broad support for the environment among upper, middle and lower socio-economic classes.

• People in cities and the country care equally about the environment—but they split over "who directs regulation and policy can likely lead to conflict." Anti-federal government attitudes in rural Idaho are more prominently expressed.

• Evidence suggests Democrats and liberals are more environmentally conscientious and that newspaper reporting on the environment is more trusted than that from television.

• Science and religion differ on issues like genetic engineering—but both groups agree on "preserving this world that we share."

It's a good message. Unfortunately, it's not being communicated in Idaho and much of the West, where environmental stewardship is constantly at odds with government policies that care more about protecting industries.

For instance, the state has not done enough to protect Idaho's groundwater, and it may be too late to reverse some of the pollution.

The message certainly has not been communicated in Washington D.C. where federal environmental protection laws are now under siege.

Despite the Catholic bishops' views, an unprecedented federal interagency study of the entire Columbia Basin—one of the most penetrating and comprehensive watershed studies ever performed on a Western watershed—is on a shelf, gathering dust.

The goals of protecting endangered species and ensuring wildlife diversity place second to enhancing special interests. For instance, recovering Idaho's endangered salmon has been stalled by bureaucratic machinations.

The politics of the day—religious leaders have learned—too often miss the connection between the message of the creation and responsible caretaking.

Reprinted by permission of the Idaho Falls Post Register

SECTION II

Origins and Roots: A Crash Course in Theological and Ethical Perspectives on Biodiversity

Biodiversity, Theology, and Ethics: Key Concepts

by Peter W. Bakken

Introduction

People talk about biodiversity in many different ways. Understanding how different perspectives shape the way an issue is discussed can help clarify what is really at stake in our public arguments and can help us better grasp the motivations and reasoning of our opponents, our allies, and ourselves. Theology and ethics offer powerful and different ways of thinking and speaking about biodiversity, yet people often confuse the two perspectives. This essay explores the ways we talk about biodiversity and the relationship between theology and ethics.

The Many Languages of Biodiversity Debates

When people discuss and debate biodiversity issues, they often seem to talk past each other. Why? This happens because each person is speaking a different "language"—using a different form of discourse. These may include:

- *Scientific discourse*—such as definitions of biodiversity, estimates of numbers of species and rates of extinction, theories about the causes and consequences of biodiversity loss;
- *Economic discourse*—measuring in dollars the costs and benefits of species protection or biodiversity loss;
- *Legal argumentation*—the relevance of particular laws or regulations to specific cases, legal jurisdictions of different governmental authorities, legal rights and responsibilities of the various parties involved;
- *Personal narratives*—emotionally powerful encounters with endangered creatures, conflicts with environmental regulations, struggles to protect habitat;
- *Expressions of values and ideals*—such as love and respect for wild animals; concerns for loggers, farmers, fisher folk, etc., and their families; hopes for a world in which human and nonhuman creatures harmoniously coexist;
- *Ethical reasoning*—applying moral rules or principles to particular situations, weighing conflicting values and interests, formulating rights for humans or nonhuman beings, defining the ethical obligations of individuals and groups for protecting species and habitat, assigning moral praise or blame;
- *Religious or quasi-religious affirmations*—stories, poetry, doctrines, beliefs, codes of conduct drawn from particular historical religious traditions; spiritual or philosophical understandings of the fundamental character of reality, ultimate values, the nature of human beings, and the purpose of existence or the meaning of life.

The Role of Religious/Theological Discourse

For some people, the use of religious language in public debates is especially problematic. They are afraid that it will be a "conversation-stopper," invoking an unquestionable ultimate authority to back a position and place it beyond any possibility of reasoned discussion: "God said it, I believe it, that settles it," as one bumper sticker proclaims. But religious language can also help expand the boundaries of discussion by introducing considerations of value, meaning, and worldview, in addition to those of science, economics, and law. Sometimes in discussions about biodiversity such language may contribute to opening the discussion beyond narrow technical

KEY POINTS

Different kinds of language—or discourse—can add new ways of understanding or communicating about biodiversity.

Theology and ethics can add dimensions of value and meaning to otherwise narrow and technical debates about biodiversity.

considerations by drawing attention to neglected dimensions of the issue. Religious language can help "smoke out" hidden value and worldview assumptions hiding beneath purportedly "neutral" scientific or economic analysis and clarify what is most profoundly at stake in the issue for the parties involved.[2]

Religious advocates for biodiversity preservation intend more than merely procedural clarification, of course. They hope that, by highlighting the religious dimensions of the issue, they can help some of their opponents to realize that their own deepest beliefs may be at odds with the actions or policies that they themselves are promoting.

Theology and Ethics: Differences and Connections

In environmental discussions, questions of ethics, attitudes, morality, worldview, spirituality, religion, and cultural norms are often lumped together under the heading "values." When values enter the debate, the terms 'theology' and 'ethics' emerge and typically are used interchangeably. While similarities exist between the two disciplines, there are also important differences.

Differences
Theology can be defined as reflection on the teachings, symbols, stories, and practices of a religious tradition. Sometimes such reflection relates these elements of the tradition to the meaning and transcendence that are present in areas of human experience—science, art, interpersonal relationships, encounters with the natural world—that are not usually labeled "religious" or may not involve an explicit notion of the "supernatural." Theological reflection can take many different forms, from the framing of carefully reasoned, systematic arguments to the articulation of poetic visions or personal narratives that seek to evoke rather than demonstrate their truths. But theology always involves *thinking about* the raw materials provided by a religious tradition and relating them to each other and to other realms of human thought and experience.

Ethics can be defined as reflection on the rules for behavior, principles of action, ways of life, ideals for individuals or societies, etc. that *ought* to govern relationships among human beings and between human beings and other creatures. Like psychology, sociology, and anthropology, ethics examines human attitudes and behavior, but it does so in order not merely to *describe* them, but to *prescribe* them; to say how people should behave, not simply how they *do* behave. As a form of reflection, it does not simply state the rules, ideals, norms, etc. that it commends, but it seeks to justify them, apply them to problematic cases, and resolve conflicts between them.

Connections
How, then, are theology and ethics related? Their connections are complex, varied, and subject to seemingly endless argumentation, but here I will highlight three important ways that theology relates to ethics.

• *Historical Source.* Religious traditions involve the "handing down" of teachings, stories, symbols, and practices within a community from one generation to the next. Implicit in that legacy are behaviors that the tradition defines as obligatory or praiseworthy. That historical legacy can have a powerful influence, even on those who no longer accept some or all of the religion's precepts. Medieval historian

* Religious language can also help expand the boundaries of discussion by introducing considerations of value, meaning, and worldview, in addition to those of science, economics, and law.

ORIGINS AND ROOTS: A CRASH COURSE IN THEOLOGICAL AND ETHICAL PERSPECTIVES ON BIODIVERSITY

COMMUNICATIONS TIP

The many ways of talking about biodiversity (scientific, economic, legal, ethical, etc.) are related to each other, often in complex and subtle ways, but it is important to recognize the differences among them and to notice when someone (including oneself) is shifting from one type of language to another.[1]

Lynn White, Jr., for example, argued that Western civilization retains the impress of its Christian origins in its assumption that all things exist for the use and benefit of human beings.[3] Others have seen typically Christian ethics in the activities and philosophies of "secular" environmentalists.[4]

• *Grounds for Arguments.* For adherents of a religious tradition, that tradition's teachings can serve as logical warrants for the norms on which they base their ethical reasoning, authorizing those norms and embedding them in a belief that "this is the way things really and ultimately are." Those ethical norms can be shared with others outside the tradition, even if, for the believer, they are ultimately based on particular religious beliefs. For example, a believer may regard respect for human beings as grounded in the religious teaching that "humans are made in the image of God," or that nature is to be respected because it is "God's creation." Others with different religious beliefs, or no religious beliefs, may share such respect for persons or nature but ground it in some other way.

• *Answers to Moral Questions.* Religious or theological language can be used to answer questions that are raised by the effort to live a moral life or to make ethical decisions. Why be ethical? What are we doing when we are being ethical? Why can it be so hard to behave ethically? What do we do with moral failure—ours or that of other persons? What ultimately will come of our best moral efforts, given their seeming fragility in the face of the forces working against them? These questions can, of course, be answered in non-religious ways as well, but answering them is one of the jobs that theology is often called upon to do.

Notes

[1] For further discussions of different forms of discourse involved in policy debates, see Roger Hutchinson, *Prophets, Pastors and Public Choices: Canadian Churches and the Mackenzie Valley Pipeline Debate* (Waterloo, Ontario: Wilfrid Laurier University Press, 1992); and James M. Gustafson, "Varieties of Moral Discourse: Prophetic, Narrative, Ethical and Policy," in Calvin College and Calvin Theological Seminary, *Seeking Understanding: The Stob Lectures 1986-1998* (Grand Rapids, MI: Eerdmans, 2001).

[2] Nicholas Wolterstorff distinguishes between the role of religious belief in "acceptance governance"—controlling whether or not we accept a certain proposition—and its role in "direction governance"—controlling what things we think about. See his *Until Justice and Peace Embrace* (Grand Rapids, MI: Eerdmans, 1983), 168-69.

[3] Lynn White, Jr., "The Historical Roots of Our Ecological Crisis," *Science* 155 (10 March 1967): 1203-4. For just one (among many) counterarguments, see Elspeth Whitney, "Ecotheology and History," *Environmental Ethics* 15 (Summer 1993): 151-69.

[4] See, for example, Linda Graber, *Wilderness as Sacred Space*, Monograph Series of the Association of American Geographers (Washington, D.C.: Association of American Geographers, 1976); Donald Worster, "John Muir and the Roots of American Environmentalism," in *The Wealth of Nature: Environmental History and the Ecological Imagination* (New York: Oxford University Press, 1993), 184-202; and Mark Stoll, *Protestantism, Capitalism, and Nature in America* (Albuquerque, NM: University of New Mexico Press, 1997).

ETHICS IN THE MEDIA

Redwood Rabbis

by Seth Zuckerman
SIERRA Magazine
November/December 1998

It was a ritual at once traditional and radical that drew 250 people to an ancient redwood grove ten miles from Northern California's Headwaters Forest on a stormy January day in 1997. Between rain squalls they were celebrating Tu B'shvat—the Jewish Festival of Trees. But this ceremony was not just about spiritual connection with the plant kingdom, and included more than the usual ritual meal of fruits, nuts, and wine. The forestry chair of the local Sierra Club chapter gave an overview of the threat posed to the old-growth redwood forests by the Houston-based Maxxam Corporation. Another worshipper chanted the haunting Kaddish, or mourner's prayer, in memory of creatures displaced or killed by logging. Most radical of all, the ceremony set the stage for an act of civil disobedience: the planting of redwood seedlings on an eroding stream bank on Maxxam property to symbolize hope for the restoration of land already clear-cut and creeks stripped of their tree cover. Maxxam had refused permission to plant, but the worshippers vowed they would break the law and trespass, seedlings and shovels in hand.

The religious action was part of a larger campaign to invoke Jewish traditions in defense of Headwaters Forest, the largest tract of unprotected ancient redwoods in the world, acquired by Maxxam in a hostile takeover of Pacific Lumber Company in 1986. Because Maxxam CEO Charles Hurwitz is a leading member of Houston's Jewish community, organizers have been seeking to appeal to him by contrasting his actions with Jewish teaching. They're also working to build a strong Jewish constituency for the protection of old-growth redwoods and other ecosystems, a campaign that's part of a nationwide interfaith effort to apply spiritual principles in environmental battles.

Such applications are hardly new—the Book of Deuteronomy, for example, prohibited the Israelites from destroying the fruit trees of cities they besieged. Activists tapped this tradition in 1995 by sending a letter to Hurwitz just before Yom Kippur, the Day of Atonement, when observant Jews reflect on their actions of the preceding year.

The invocation of Jewish values may have touched a nerve at the top of Maxxam. At an interfaith press conference on Headwaters in the spring of 1996 in nearby Eureka, Rabbi Lester Scharnberg wondered aloud whether "perhaps Mr.

ORIGINS AND ROOTS: A CRASH COURSE IN THEOLOGICAL AND ETHICAL PERSPECTIVES ON BIODIVERSITY

Hurwitz has forgotten the faith of his ancestors." Scharnberg's remarks, carried on the wire services and picked up by the Houston press, drew a stinging phone call from Hurwitz's rabbi, Samuel Karff, who disputed whether this member of his congregation deserved rebuke. Karff defended Hurwitz as a charitable man; the Hurwitz family has donated heavily to Karff's Temple Beth Israel, and the synagogue's school is housed in the Hurwitz Building.

Despite their disagreement, Karff arranged for Scharnberg to speak with Hurwitz directly. In the 45-minute conversation that ensued, Hurwitz was taken aback to find a rabbi on the other side of the Headwaters battle, recalls Scharnberg. "He didn't know me, but he has an image of what a rabbi is," Scharnberg says, "and he expressed surprise that I was aligned with 'conga drums, dreadlocks, tie-dye, and hippie radicals who threaten to kill, maim,' and so forth. I said, 'I'm not aligning myself with people who kill, but I am an environmentalist.'"

Scharnberg didn't have an opportunity to confront Hurwitz again until the May 1998 Maxxam stockholders meeting, armed with a proxy signed over to him by another Headwaters activist. Christian and secular speakers addressed issues of science, economics, and corporate responsibility, and left religion up to Scharnberg. That was probably a wise call, given that the roster of Maxxam's officers and board members has a substantial Jewish representation.

Scharnberg asked the board if Maxxam had considered moral questions in the course of its operations, and, if not, how the firm could hope to act ethically. The very question provoked a firestorm of response that continued after the 90-minute official meeting. "The directors of Maxxam were outraged that we should introduce religion into this board meeting," Scharnberg says.

[That] summer, the Coalition on the Environment and Jewish Life—which claims such prominent member groups as Hadassah, Hillel, B'nai Brith, and the American Jewish Congress—called for stronger habitat protections in Headwaters and in all remaining old-growth redwood groves. Several other major Jewish organizations have adopted or are considering similar resolutions. And on Hurwitz's home turf, a group of Houston Jews rented the Jewish Community Center for another ecologically oriented Tu B'shevat.

Back in the redwoods in January 1997, a caravan of 100 worshippers—some wearing talliths, or fringed prayer shawls, as Jews have for thousands of years—hiked onto the timber firm's property and planted two dozen redwood seedlings along a barren stream bank. Some used shovels, some trowels, some their bare hands. Longtime Earth First! activist Darryl Cherney described it as a miracle. "At a place where demonstrators before have been met with billy clubs, nightsticks, and arrests, we are now walking freely," he said. "It reminds me of the parting of the Red Sea." Nearly two years have passed, and student rabbi Steinberg—who lives just a few miles away—hasn't revisited the site. "I'd rather remember the trees beautifully planted than to see that Pacific Lumber has pulled them up or that the whole bank has fallen away," she says.

Steinberg reminds activists to look at the big picture. "If you approach a campaign like this as spiritual work, the moments along the way can be transformative to you as an individual soul." It's that transformation of souls that will determine whether "the forest trees shout for joy," as the Psalmist sang.

PHOTO BY LYNN BETTS, USDA NATURAL RESOURCES CONSERVATION SERVICE

Reprinted with permission of Seth Zuckerman. Seth Zuckerman is co-author of Salmon Nation: People and Fish at the Edge. *More of his work can be seen at www.tidepool.org/dispatches.*

Overview of World Religions

by Jane Elder and Marian Farrior

Introduction

Throughout human culture, religious and spiritual beliefs have often framed the way people value and treat biodiversity and nature. In recent years, theological teachings on the environment have become a focal point, not only for scholars, but also for activists from both secular and religious perspectives.

Any attempt to briefly summarize and categorize the vast array of religious traditions found around the world will necessarily be a crude oversimplification. The following three categories, however, may prove useful in providing an overview of religious thought on the natural world and on biodiversity.

Earth-based, Animistic, and Pantheistic Religions

Many indigenous peoples and traditional societies practice spiritual traditions in which living nature, from an individual bird to a sacred mountain, has its own spirit and spiritual power, or deific quality. The relationship with nature is therefore one of kinship, as opposed to management or oversight. Many, if not most, of these spiritual traditions are not based in written texts, but in story, oral tradition, the wisdom of elders, and ritual. The romanticized view of traditional cultures is one of bucolic harmony with nature across the generations. However, although some cultures have successfully sustained the environment for eons, others have exploited nature in ways that are ecologically non-sustainable.

Many indigenous conservation practices include an ethos providing total protection to some habitats (such as sacred groves or springs). Akin to the practices of Eastern and monotheistic religions, these cultural practices often provide total protection to specific species (certain plants or animals are considered taboo), they protect wildlife during critical life stages (such as nesting stages), and they mandate local stewards to supervise resource use.[1]

Asian and Eastern Religions

The Eastern religions have common threads that link aspects of spiritual practice to harmony with nature and its forces through heightened awareness and ritual. For example, in Taosim, one strives to be in harmony with the Tao, "the way" of being that governs all things.

Instead of a single god, many of these traditions embrace a multiplicity of gods and spirits endowed with various roles and powers (polytheism). In both Buddhism and Hinduism, living in harmony with nature and showing respect for living beings have great significance. For example, the Buddhist Vietnamese monk Thich Nhat Hahn has drafted "Fourteen Precepts for the Order of Interbeing" on ethical relationships that can be directly applied to environmental issues.[2] Despite these teachings of harmony, however, Asian cultures have contributed their share of deforestation, species loss, and degradation of habitat.

Monotheistic Religions

The dominant monotheistic (believing in one god) religions—Judaism, Christianity, and Islam—are also known as the Abrahamic religions, since the roots of each can be traced back through the lineage of the biblical Abraham. Within the monotheistic traditions, humans have a distinct spiritual significance as being created in the image

KEY POINTS

Spiritual traditions often inform the way people understand and act toward the environment.

The diversity of beliefs among the world's religions presents opportunities to frame common solutions to the biodiversity crisis according to the unique perspectives of each faith.

of God. In these traditions, this status has sometimes been used to assign "higher" value and privilege to humans than to the rest of nature. However, in these same traditions, humans carry significant responsibilities as stewards and caretakers of the rest of creation. The tensions between dominion (human primacy) and stewardship (human responsibility) pose some of the more challenging ethical conundrums for cultures rooted in monotheistic traditions. Within these traditions, the weighing of human versus nonhuman interests and the scope of human responsibility for the natural world have been questions of the ages. Environmental harm may be attributable to the human tendency to see ourselves as most important, while the religious teachings hold that humans should observe the primacy of God. These historic tensions have been interpreted and exploited to serve many ends beyond religious practice.

The considerable distinctions among other basic beliefs of these religions, and the widely divergent interpretations and practices within these three religions, make it extremely risky to draw broader generalizations about applied environmental ethics.

Toward Common Ground and Appreciation for Diversity

The richness and diversity of human spiritual practice throughout history and around the globe present challenges for finding common ground that can help resolve ethical questions for biodiversity conservation, especially where solutions will require cooperation across nations and cultures. But there is also opportunity.

In 1986, an interfaith gathering of leaders in Buddhism, Christianity, Hinduism, Islam, and Judaism met at the Basilica of St. Francis in Assisi, Italy, to state their environmental beliefs. Organized by the World Wildlife Fund (WWF) and the Network on Conservation and Religion, this historic occasion yielded the Assisi Declarations, which are statements of spiritual beliefs and ethical responsibilities regarding nature. Since that time, three more religions, Baha'i, Jain, and Sikh, have added their declarations, and the Alliance of Religion and Conservation has been formed to work with the WWF to preserve

Resources on Religion and Nature
For a deeper discussion on this topic, see:

- Forum on Religion and Ecology (http://environment.harvard.edu/religion/)
- Encyclopedia of Religion and Nature (www.religionandnature.com/)
- United Nations Environment Programme's book *Cultural and Spiritual Values of Biodiversity*, particularly the World Religions Bibliography in the Resources section.

Additional contacts and resources include:
- The National Religious Partnership for the Environment (www.nrpe.org)
- The Coalition on the Environment and Jewish Life (www.coejl.org)
- The Web of Creation (www.webofcreation.org)
- The Biodiversity Project's handbook, *Building Partnerships with the Faith Community: A Resource Guide for Environmental Groups* (www.biodiversityproject.org)
- Beliefnet.com (www.beliefnet.com)
- Adherents.com (www.adherents.com)

See the Resources section at the end of this Handbook for contact information for other organizations working at the intersection of spirituality, ethics, and the environment.

precious habitat. A text of the Assisi Declarations can be found in Appendix I.

The Earth Charter, discussed in Section IV, is another example of how religious and spiritual leaders from throughout the world have worked together to find the common base of values that define an "Earth ethic" for human and ecological sustainability. The Earth Charter was drafted in response to the United Nation's Rio Earth Summit and the urgency of global environmental problems. Working with great care for the past ten years to respect a wide array of spiritual practices and traditions, the crafters of the Earth Charter have achieved something remarkable—a shared statement of human spiritual values as they apply to protecting and sustaining the Earth's environment. It speaks to creating a global-scale ethic, addressing social and economic justice, and maintaining ecological integrity.

The diversity of religious traditions and approaches, combined with a unifying global ethic, is a powerful tool for social transformation. As stated in the United Nation's Environment Programme's seminal work, *Cultural and Spiritual Values of Biodiversity*, "If biological diversity is good for the environment, maybe conceptual and practical, cultural and religious diversity and interaction is good too."[3]

Notes

[1] Fikret Berkes, "Religious Traditions and Biodiversity," *Encyclopedia of Biodiversity 5* (San Diego, CA: Academic Press, 2001): 117-18.

[2] Joan Halifax and Marty Peale, "Interbeing: Precepts and Practices of an Applied Ecology," in Darrell Addison Posey et al., eds., *Cultural and Spiritual Values of Biodiversity* (London, UK: Intermediate Technology Publications, United Nations Environment Programme, 1999), 475-80.

[3] Tim Jensen, "Forming 'the Alliance of Religions and Conservation,'" in Darrell Addison Posey et al., eds., *Cultural and Spiritual Values of Biodiversity* (London, UK: Intermediate Technology Publications, United Nations Environment Programme, 1999), 499.

Major Religions of the World
Ranked by Number of Adherents

Last modified 25 March 2002. *(Sizes shown are approximate estimates and are here mainly for the purpose of ordering the groups, not providing a definitive number. This list is sociological/statistical in perspective.)*

Christianity: 2 billion
Islam: 1.3 billion
Hinduism: 900 million
Secular/Nonreligious/Agnostic/Atheist: 850 million
Buddhism: 360 million
Chinese traditional religion: 225 million
primal-indigenous: 150 million
African Traditional & Diasporic: 95 million
Sikhism: 23 million
Juche: 19 million
Spiritism: 14 million
Judaism: 14 million
Baha'i: 6 million
Jainism: 4 million
Shinto: 4 million
Cao Dai: 3 million
Tenrikyo: 2.4 million
Neo-Paganism: 1 million
Unitarian-Universalism: 800 thousand
Scientology: 750 thousand
Rastafarianism: 700 thousand
Zoroastrianism: 150 thousand

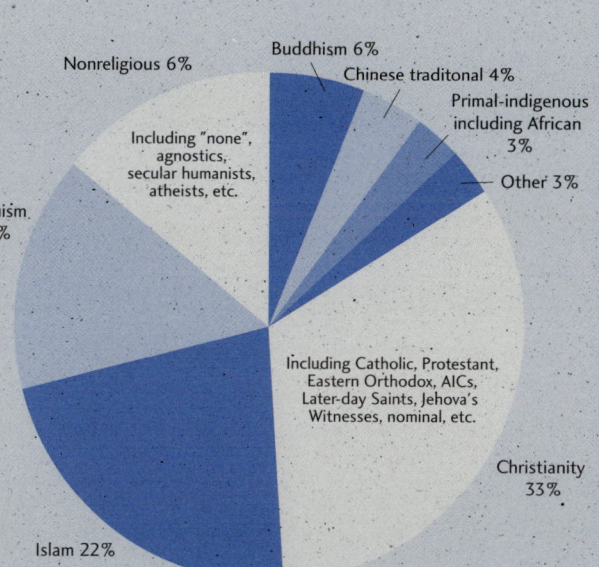

The adherent counts presented in the list above are estimates of the number of people who have at least a minimal level of self-identification as adherents of the religion. Levels of participation vary within all groups. These numbers tend toward the high end of reasonable worldwide estimates. Valid arguments can be made for different figures, but if the same criteria are used for all groups, the relative order should be the same.

This listing is not a comprehensive list of all religions, only the "major" ones. There are distinct religions other than the ones listed above. But this list accounts for the religions of over 98% of the world's population. Statistics and geography citations for religions not on this list, as well as subgroups within these religions (such as Catholics, Protestants, Karaites, Wiccans, Shiites, etc.) can be found in the main Adherents.com database.

© www.adherents.com

Biblical and Theological Perspectives on Biodiversity

Peter W. Bakken
(with contributions from Daniel Swartz)

Introduction

Public opinion research affirms that the spiritual connection to biodiversity is important in American culture; it is a source of the ethical justifications for protecting biodiversity for itself and for its benefits to humans.

Biblical religious traditions have profoundly influenced the philosophical, cultural, and political heritage of the United States, including many of the leading arguments for and against protecting biodiversity. Today, 77% of Americans identify themselves as Christian or Jewish, while 13% identify themselves as secular or nonreligious.[1] Understanding the theological underpinnings of the dominant American cultural views on the human relationship with nature is an essential part of advocating biodiversity protection.

Biblical Principles for Protecting Biodiversity

The biblical texts include several core principles defining how humans ought to relate to God's creation:
- Creation is good;
- The great variety of life is intentional, and the myriad creatures fulfill God's purpose—every creature has a place in creation, and creation in its awesome variety is a source of wonder and joy;
- Humans have been charged with "keeping the garden"—taking care of creation and keeping God's covenant with all creation.

Creation is Good

In the story of creation recounted in Genesis, the phrase "of every kind" is repeated and echoed (e.g., "every x after its kind")—in connection with plants and fruit trees (Gen. 1:11-12), sea creatures and birds (Gen. 1:20-21), and land animals (Gen. 1:24-25).[2] The fact that creation is comprised of distinct yet interconnected parts is also underscored by the very structure of the narrative (divided into distinct days, each involving a specific domain or element of the universe), and the various acts of separation: light from darkness (Gen. 1:4), the waters under the firmament from the waters above (Gen. 1:7), land from seas (Gen. 1:9), and day from night (Gen. 1:14). Finally, the various parts of

KEY POINTS

The Bible and its theological interpretations have influenced the way many Americans view biodiversity.

Biblical passages affirm that Creation is good and a blessing, that the diversity of life on Earth is intentional and is in accordance with God's purposes, and that humans have an obligation to act responsibly toward it.

At the same time, Biblical passages have also been used to support less salutary responses to the environment, including ideas that nature is corrupt or imperfect, that humans have a right to dominate other species, and that the world is coming to an end and therefore not worthy of concern.

Interpretations of the Bible or other sacred religious texts can be used to justify almost any action toward the environment. But the multitude of passages that emphasize the importance of environmental stewardship can help spiritually motivated biodiversity advocates respond to the views of those who use the Bible to justify the exploitation of nature and open dialogues with other believers.

creation are seen by God to be *"good,"* and the whole comprising them is *"very good"* (Gen. 1:31, emphasis added). Diversity as such is an integral element in the "very good" creation as a whole.

Furthermore, many have argued that as a creature of God, each being is entitled to care and respect. Living creatures are the objects of God's providential care, which insures the survival of their "kinds." God provides them with the ability to reproduce themselves—the trees and plants are provided with seeds (Gen. 1:11-12) and animals are given the blessing to "be fruitful and multiply" (birds and sea creatures, Gen. 1:22; humans, Gen. 1:28)—and with the gift of food (in the form of "green plants," Gen. 1:29-30).

The relationship of creature to habitat is emphasized by the narrative structure, in which the varied spaces of creation are created first (day and night, sky and ocean, land) and then the corresponding inhabitants (sun, moon, stars; birds and sea creatures; land animals and humans). There is a place for everything, and everything in its place. The theme of interrelatedness continues in the second chapter of Genesis, which gradually builds up a network of relationships within creation: human being to land, animal to human, and human to human (Gen. 2:4-24).[3]

Finally, part of the world's goodness takes the form of the material and spiritual sustenance it provides to humans. The early rabbinic commentary on Genesis, B'reishit Rabbah, quotes Rabbi Levi Ben Hiyyata as teaching that "without earth, there is no rain, and without rain, the earth cannot endure, and without either, humans cannot exist." Midrash Tehillim, a medieval commentary on the Psalms, states that rain is a greater blessing than the giving of the ten commandments, for the commandments bless those who abide by them while rain blesses the entire world. But the world's gifts go beyond ensuring survival. Its sheer abundant goodness—biblical language for rich biodiversity—is supposed to inspire in us abundant love and generosity. When we participate in the grander chorus of creation, we find ourselves at home, at one, whole. Thus, the Bible portrays the age of redemption, of ultimate wholeness, as a time when all people will be in harmony with all of creation: "In that day, I will make a covenant for you with the beasts of the field, the birds of the air, and the creeping things of the ground. I will banish bow, sword and war from the land and let them lie down in safety" (Hos. 2:20).

The Great Variety of Life Is Intentional: "How manifold Are Thy works!" Psalms 104 and 148

Psalm 104 recapitulates many of the same themes. It recalls Genesis 1 but speaks in the present tense and with greater attention to detail, as it presents a variety of creatures (water, grass, goats, badgers, cedars, fir trees, lions, storks, leviathan, human beings) and their ecological interrelationships. The sheer multiplicity of kinds of things inspires wonder in the Psalmist: "O Lord, how manifold are thy works! In wisdom hast thou made them all; the earth is full of thy creatures. Yonder is the sea, great and wide, which teems with things innumerable, living things both small and great" (Ps. 104:24-25).

✹ *Understanding the theological underpinnings of the dominant American cultural views on the human relationship with nature is an essential part of advocating biodiversity protection.*

ORIGINS AND ROOTS: A CRASH COURSE IN THEOLOGICAL AND ETHICAL PERSPECTIVES ON BIODIVERSITY

Psalm 104

(From the King James version of the Bible)

1: Bless the LORD, O my soul. O LORD my God, thou art very great; thou art clothed with honour and majesty.

2: Who coverest thyself with light as with a garment: who stretchest out the heavens like a curtain:

3: Who layeth the beams of his chambers in the waters: who maketh the clouds his chariot: who walketh upon the wings of the wind:

4: Who maketh his angels spirits; his ministers a flaming fire:

5: Who laid the foundations of the earth, that it should not be removed for ever.

6: Thou coveredst it with the deep as with a garment: the waters stood above the mountains.

7: At thy rebuke they fled; at the voice of thy thunder they hasted away.

8: They go up by the mountains; they go down by the valleys unto the place which thou hast founded for them.

9: Thou hast set a bound that they may not pass over; that they turn not again to cover the earth.

10: He sendeth the springs into the valleys, which run among the hills.

11: They give drink to every beast of the field: the wild asses quench their thirst.

12: By them shall the fowls of the heaven have their habitation, which sing among the branches.

13: He watereth the hills from his chambers: the earth is satisfied with the fruit of thy works.

14: He causeth the grass to grow for the cattle, and herb for the service of man: that he may bring forth food out of the earth;

15: And wine that maketh glad the heart of man, and oil to make his face to shine, and bread which strengtheneth man's heart.

16: The trees of the LORD are full of sap; the cedars of Lebanon, which he hath planted;

17: Where the birds make their nests: as for the stork, the fir trees are her house.

18: The high hills are a refuge for the wild goats; and the rocks for the conies.

19: He appointed the moon for seasons: the sun knoweth his going down.

20: Thou makest darkness, and it is night: wherein all the beasts of the forest do creep forth.

21: The young lions roar after their prey, and seek their meat from God.

22: The sun ariseth, they gather themselves together, and lay them down in their dens.

23: Man goeth forth unto his work and to his labour until the evening.

24: O LORD, how manifold are thy works! in wisdom hast thou made them all: the earth is full of thy riches.

25: So is this great and wide sea, wherein are things creeping innumerable, both small and great beasts.

26: There go the ships: there is that leviathan, whom thou hast made to play therein.

27: These wait all upon thee; that thou mayest give them their meat in due season.

28: That thou givest them they gather: thou openest thine hand, they are filled with good.

29: Thou hidest thy face, they are troubled: thou takest away their breath, they die, and return to their dust.

30: Thou sendest forth thy spirit, they are created: and thou renewest the face of the earth.

31: The glory of the LORD shall endure for ever: the LORD shall rejoice in his works.

32: He looketh on the earth, and it trembleth: he toucheth the hills, and they smoke.

33: I will sing unto the LORD as long as I live: I will sing praise to my God while I have my being.

34: My meditation of him shall be sweet: I will be glad in the LORD.

35: Let the sinners be consumed out of the earth, and let the wicked be no more. Bless thou the LORD, O my soul. Praise ye the LORD.

Individual creatures are the objects of God's providential care: God makes "springs gush forth" that give water to animals and plants (Ps. 104:10); God makes grass grow for cattle, and grapes, olives, and grain for people (Ps.104:14-15); and God gives the "young lions . . . their prey" (Ps. 104:21). Continuity is assured, for even as God takes away creatures' breath so that they die, God's spirit (re)creates them and "renews the face of the ground" (Ps. 104:29-30).

Among the ecological relationships highlighted in this psalm is that of creature to habitat: the birds sing among the branches of the trees growing by the streams; the birds nest in the cedars, and the stork in the fir tree; the mountains are "for" the wild goats, and the rocks for the badgers (or conies) (Ps. 104:12, 16-18). Each creature has its place within creation.[4]

The value of creation's diversity for God is given a different sort of expression in Psalm 148. There, all of creation is called upon to praise God, from angels, sun, moon, and stars through sea monsters, hail and frost, fruit trees, beasts, birds, and every sort and condition of human being. It is difficult to resist the overwhelming impression that the diversity of creation itself is an occasion for praise and a source of joy to both creature and creator.[5]

Humans are Responsible for Keeping God's Covenant with Creation: Noah and the Flood

The story of Noah and the flood emphasizes the importance of creation to God and humanity's responsibility to preserve it.

In the story, all of creation—all flesh—is infected by violence, although human beings appear to be the ones primarily responsible (Gen. 6:5-6, 11-13). God thus determines to "blot out" all living things, human and nonhuman (Gen. 6:7, 13, 17). However, God decides to preserve a remnant of creation—a remnant which includes, again, both human and nonhuman creatures —by commanding Noah to build a huge ark that will ride out the flood (Gen. 6:18-20, 7:1-3). And, finally, when God resolves not to destroy the Earth in this way again, God's covenant is not only with Noah and his family, but with all the animals that were on the ark and with the Earth itself: "And God said, 'This is the sign of the covenant which I make between me and you and every living creature that is with you, for all future generations'" (Gen. 9:12). The understanding of creation in this story is emphatically holistic: Humankind and the birds and animals live or die together; neither can survive without the other. The ark itself is a kind of microcosm, a fragile community sustained in the midst of the waters of chaos.

The story also underscores the value of each individual species: Every "kind"—the "unclean" (i.e., useless to humans) as well as the "clean" (i.e., useful to humans)—is to be taken aboard the ark (Gen. 7:2-3). And during the flood, "God remembered Noah and *all the beasts and all the cattle* [i.e., the wild and the domestic animals] that were with him on the ark" (Gen. 8:1, emphasis added).

As a habitat, the ark is only a temporary dwelling. When the floodwaters recede and the ark comes to rest, God's command for the animals' debarkation echoes Genesis 1: "Bring forth with you every living thing that is with you of all flesh—birds and animals and every creeping thing that creeps upon the earth—that they may breed abundantly on the earth, and be fruitful and multiply upon the earth" (Gen. 8:17). The goal is a creation in which each creature has reclaimed its proper niche.[6]

Later Tradition

The texts we have just considered are from the Hebrew Scriptures, part of the common heritage of Judaism and Christianity. Each faith reads them, however, through intervening layers of traditions and interpretation—Judaism, through the generations of

> "[God] produced many and diverse creatures, so that what was wanting to one in representation of the divine goodness might be supplied by another... hence the whole universe together participates in the divine goodness more perfectly."
>
> *St. Thomas Aquinas*

interpretation and application of the laws and stories by rabbis (as seen above in the examples from B'reishit Rabbah and Midrash Tehillim); Christianity, through the lens of the ancient Christian scriptures (New Testament) and theological traditions, including those which, from the early church to medieval times, tried to relate, and often to synthesize or harmonize, biblical teachings with philosophical ideas from ancient Greece.

The primary contribution of the New Testament texts to Christian views on biodiversity has been the reaffirmation of creation and God's care for it through Jesus' teachings (Matt. 6:28-29: "Consider the lilies of the field, how they grow; they neither toil nor spin; yet I tell you, even Solomon in all his glory was not arrayed like one of these"); the identification of Jesus Christ, in his divine nature, with the Creator and, as incarnate in human form, with the creation (John 1:3-4: "All things were made through him, and without him was not anything made that was made. In him was life, and the life was the light of men"); and eschatological teachings about a "new heaven and a new earth" (Rev. 21:1).

One later theological development that has significance for biodiversity is that of the "principle of plenitude," which refers to the fullness of the variety of things that exist in creation. The argument is that God's infinite, overflowing goodness expresses itself in a creation that contains every possible kind of creature.[7] This argument goes back to Plato; in Christian form, it can be found in St. Thomas Aquinas, who is cited in the recent U.S. bishops' pastoral on the environment, *Renewing the Earth*:

> The diversity of life manifests God's glory. Every creature shares a bit of the divine beauty. Because the divine goodness could not be represented by one creature alone, Aquinas tells us God "produced many and diverse creatures, so that what was wanting to one in representation of the divine goodness might be supplied by another... hence the whole universe together participates in the divine goodness more perfectly, and represents it better than any single creature whatever" (*Summa Theologica*, Prima Pars, question 48 ad 2).[8]

The Lutheran theologian Joseph Sittler made a similar affirmation about the diversity of creation (although not explicitly backed by the same philosophical argument) in an article published in 1970:

> Grace comes in colors.... [This understanding of grace] is bound up with the unthinkable variety of God the Creator who loves all colors, textures, forms, nuances, and modes of life. It is grace as the joyful acknowledgment of the variety that God loves, the variety [God] has made.[9]

Against Nature: Interpretations that Devalue the Natural World

Otherworldliness: Nature as Base, Messy, and Suspect

One aspect of the Greek heritage of Christian theology has not been salutary. The Greek belief that the perfect must be "simple"—unitary, undivided, and indivisibly singular—and changeless has sometimes made it difficult for Christians to reconcile God's perfection—and the perfection sought by the Christian believer—with the messy, transitory, multifaceted character of the biophysical world.

Ambivalence toward the natural world has also been supported by a particular interpretation of Genesis 3:17, in which God responds to Adam's disobedience by declaring, "cursed is the ground because of you." Some Christians interpret this to mean that the natural world is "cursed," or (like humanity) "fallen," and no longer operates as God originally intended. Pain, disease, death, predation, competition, etc. are seen as "natural evils" resulting from creation's corruption. In its most extreme forms, this view resembles "Gnostic" teachings (which circulated in ancient

Greek and Roman culture but were eventually declared heretical by the church) that the creation of the material world was a tragic accident or the work of an inferior or malign deity.

Christians have therefore sometimes turned away from the material creation toward "other-worldliness," seeking a purely spiritual world "above" the material one, or in apocalypticism, hoping for an everlasting, purified spiritual world to follow the dissolution of the present transitory and corrupt one. In these theologies, the Earth is at best a stage or backdrop for the drama of human redemption, without real significance for God or in itself. As Katherine Hepburn's character, Rosie, in the movie *The African Queen*, states, "Nature, Mr. Aulnaut, is what God put us on Earth to rise above."

Otherworldliness has been less of a theological tendency in Judaism, as witnessed in the array of brachot (blessings) dealing with natural phenomena from mountains to unusual creatures. The historical experiences of Jews in the past millennium, however, have often reduced or eliminated their intimate contact with the land.

"Dominion" as Domination: The World Is First, and Foremost, for Human Benefit
A more recent development, which has come with the rise of the modern age (and the rise of capitalism and individualism), has been an aggressive, human-centered activism, which seeks to reshape the world for human benefit, with little or no regard for what that might mean for the welfare of other creatures. This approach is now most commonly and vehemently represented by some "conservatives," who are more concerned with maintaining the present Western trajectory of largely unrestrained economic growth and technological advancement than preserving the integrity of the created nonhuman world.

In contrast with otherworldly and apocalyptic spirituality, this aggressive anthropocentrism is more "worldly": Nature has value as raw material for the meeting of human needs. At its best, this activism is the expression of a genuine love of the (human) neighbor that takes seriously the neighbor's bodily needs as a material, biological creature, for food, clothing, shelter, and medicine. It defends human dignity, freedom, and worth by appealing to the biblical teaching that humans alone are said to be created in the image of God (Gen. 2:27).

However, this sort of activist, human-centered orientation also often overlooks the way its humanitarian and democratic goals are undermined by the environmental degradation that the pursuit of the goals, through aggressive economic and technological expansion, leaves in its wake. It also tends to minimize the moral significance of the well-being or survival of nonhuman creatures and species, as well as the damaging side-effects (for people) of unlimited growth. Even when the independent value of other creatures is acknowledged, the threshold for overriding that acknowledgement in the face of competing human interests is often set very low indeed.

An actively "managerial" approach to the natural world can take a less anthropocentric form, if the belief that nature is "fallen" is combined with care for nonhuman creatures and confidence in the ability of human beings to repair creation's flaws. This view can be used to justify human interventions in nature, such as "taming" the wilderness, genetic engineering, environmental management, protection of wild animals from predation and suffering, and so on. (A different idea, that creation is "unfinished"—good but not yet perfect—and that humanity's role is to help complete it, is also used to justify these sorts of interventions in nature.)

Apocalypticism: Leaving the Earth behind
Equally problematic is a stream of thought, most prominent in apocalyptic Christianity but present to some extent in most religions, that complete trust in God and God's

* The essence of humans' special value is primarily a matter of function rather than status. The dignity of human beings derives from their distinctive relationship to God and the world, their unique role in and on behalf of the whole—as steward, servant or priest of creation.

providence means not worrying about the future. In its most extreme form, the future is utterly discontinuous with the present, so the present is completely devalued—its destruction viewed as inconsequential or even good, hastening the age of redemption. From an environmental standpoint, such beliefs became most notorious when James Watt, during his term as Secretary of the Interior, supposedly argued that conservation was unnecessary because Christ would be returning soon to bring history to a close, and that using up the earth was only to the good. (This is actually a mischaracterization of Watt's religious beliefs; in any event, his intensely pro-development policies appear to have derived more from his political than his religious views.)[10]

Responses from the Perspective of Caring for Creation

Jews and Christians who have great concern for preserving biodiversity (or other environmental objectives) have argued with those who devalue nature on scriptural ground. They have responded to otherworldliness, aggressive anthropocentrism, and apocalypticism by affirming the biblical basis for the value of biodiversity (and other environmental values)—and from that basis they have asserted that humans are bound to respect the natural world and preserve the diversity of God's creation.

To those who seek to rise above the chaos of creation, Christians have stressed the New Testament's affirmations that the world, this world, is in fact God's creation and the creation over which Jesus Christ is Lord. (Colossians 1:15-16 reads: "He is the image of the invisible God, the first-born of all creation; for in him all things were created, in heaven and on earth, visible and invisible . . . all things were created through him and for him.") Otherworldliness has been less of a temptation in Judaism.

In response to the dominionists, Christians and Jews have affirmed that even though humans do have a special value, this does not mean that every human want or need completely overrides the value of other creatures. Compromise and even self-sacrifice may be called for on the human side in some cases of conflict between human and nonhuman interests. Moreover, they have proposed that the essence of humans' special value is primarily a matter of function rather than status. The dignity of human beings derives from their distinctive relationship to God and the world, their unique role in and on behalf of the whole—as steward, servant,

---

An Ethic of Creation Care: Earthkeeping, Fruitfulness, and Sabbath

In contrast to the dominionists, who contend that nature is created solely for humanity's benefit and therefore humans have the right to exploit it in the manner we choose, Calvin B. DeWitt, an environmental scholar and leader in the evangelical environmental movement, argues that the Bible emphasizes the importance of ecological stewardship. The core of this teaching is embodied in three principles: "earthkeeping," fruitfulness, and Sabbath.

Earthkeeping comes from God's charge to the first human being to "keep" *(shamar)* the garden (Gen. 2:15). The Hebrew word *shamar,* when applied to keeping the garden "means that the Earth is to be kept by human beings in its wholesome fullness—with its vibrant, lively, dynamic character intact."[12]

The notion of fruitfulness comes from the Genesis 1 command to "Be fruitful and multiply" (Gen. 1:22, 28). But DeWitt asks, "How can human beings interpret this to mean that their blessing is given at the expense of the [parallel] blessing to the other creatures?" He argues instead that "Humans are expected to enjoy Creation and its fruits, not to destroy the fruitfulness upon which Creation's fullness depends."

The Sabbath concept extends beyond keeping the seventh day of the week holy, to resting the land from human service every seven years: "For six years you shall sow your land and gather in its yield; but the seventh year you shall let it rest and lie fallow" (Exod. 23:10-11). DeWitt states: "The Sabbath principle thus allows people to make use of the land and its creatures, but not beyond the point of wholesome sustainability."

---

38 **Biodiversity Project** *Ethics for a Small Planet: A Communications Handbook*

or priest of creation.[11]

While apocalyptic thinking does play a part in many religions, including Christianity and Judaism, the original prophetic texts typically find enough continuity between present and future times to encourage active stewardship. In response to those who use apocalyptic writings to justify a lack of concern for the Earth, many have argued that the biblical promise is for the renewal of creation, not its abolition, and that nature is to be redeemed along with people. (Isaiah 65:18 reads, "But be glad and rejoice for ever in that which I create.") Even the book of Revelation, referred to by those supporting apocalyptic views, praises God's role as creator of the world (see, e.g., 3:14, 5:13, 10:6, 14:6-7) and says that God's judgment involves "destroying those who destroy the earth" (11:18). Perhaps the most profound explication of the balance between yearning for future redemption and acting to preserve the present lies in the seemingly simple statement of Rabbi Yohanan ben Zakkai, who lived at the same time as Jesus. He wrote, "if you have a sapling in your hand, and someone should say to you that the Messiah has come, stay and complete the planting, and then go to greet the Messiah" (Avot de Rabbi Nathan, 31b). Those who care about the environment should also note that apocalyptic thinking is only one of many streams of religious thought. Furthermore, far from singling out environmental issues, in its most extreme forms of devaluing the present, apocalypticism is in fact at odds with a number of other deeply held religious values—justice, peace, and the value of human life to name three.

Sorting Conflicting Interpretations

People can legitimately disagree about the proper interpretation of the Bible and other sacred texts. As with most religious documents, verses can be found in the Bible to justify almost any action toward the environment—although some interpretations and applications may be more defensible than others. The multitude of passages supporting environmental stewardship can help spiritually motivated biodiversity advocates respond to those who use the Bible to justify the exploitation of nature, as well as open dialogues with those looking for a spiritual basis for environmental protection.

Notes

[1] From www.adherents.com.

[2] Biblical quotations are taken from the Revised Standard Version.

[3] See also Bruce C. Birch, *Let Justice Roll Down: The Old Testament, Ethics, and Christian Life* (Louisville, KY: Westminster/John Knox Press, 1991), 71-104.

[4] See also Odil Hannes Steck, *World and Environment* (Nashville, TN: Abingdon Press, 1980), 78-89.

[5] See also Claus Westermann, *The Living Psalms* (Grand Rapids, MI: Eerdmans, 1984), 255-60.

[6] Further ecologically informed readings of the story of Noah can be found in Bernhard W. Anderson, "Relation between the Human and Nonhuman Creation in the Biblical Primeval History" and "Creation and the Noachic Covenant" in his *From Creation to New Creation: Old Testament Perspectives* (Minneapolis, MN: Fortress Press, 1994), and Calvin B. DeWitt, "The Price of Gopher Wood," Faculty Dialogue (Fall 1989), 59-62.

[7] Arthur O. Lovejoy, *The Great Chain of Being: A Study of the History of an Idea* (Cambridge, MA: Harvard University Press, 1936).

[8] United States Catholic Conference, *Renewing the Earth: An Invitation to Reflection and Action on Environment in Light of Catholic Social Teaching* (14 November, 1991, U.S. Bishops' Statement), sec. III B.

[9] Joseph Sittler, "Ecological Commitment as Theological Responsibility," in Steven Bouma-Prediger and Peter Bakken, eds., *Evocations of Grace: Writings on Ecology, Theology and Ethics* (Grand Rapids, MI: Eerdmans, 2000), 86.

[10] See David Douglas, "God, the World, and James Watt," *Christianity and Crisis* 41 (1981): 258, 269-70, and Susan Power Bratton, "The Eco-Theology of James Watt," *Environmental Ethics* 6, no. 3 (1984): 195-209.

[11] For an example of this sort of interpretation of the human vocation in creation, see Douglas John Hall, *Imaging God: Dominion as Stewardship* (Grand Rapids, MI: Eerdmans, 1986).

[12] Calvin B., DeWitt, "Biodiversity and the Bible," *Global Biodiversity* 6, no. 4 (1997): 13-16.

ETHICS IN THE MEDIA

Between Heaven and Earth: Resolving Conflict in the Chesapeake Bay

by Susan Drake Emmerich
Au Sable Institute Newsletter
Winter 1999

The 650 watermen (an old English term referring to one who fishes crabs and oysters) of Tangier Island, Va., in Chesapeake Bay trace their ancestry back to Cornwall England and, because of their remote location, still speak with an Elizabethan accent. The church is the center of community life, and 80% of the people consider themselves conservative evangelical Christians.

Tangier's economy is based almost entirely on the blue crab fishery, which government officials say is suffering from over-harvesting, too much fishing gear in the water and pollution from farms and urban areas. Proposed fishing regulations led to bitter conflict between environmentalists and watermen.

The Evangelical Environmental Declaration calls for followers of Christ "...to work for the reconciliation of all people in Christ, and to extend Christ's healing to suffering Creation. God's purpose in Christ is to heal and bring to wholeness not only persons, but the entire created order."

I spent three years with the watermen in an attempt to put those principles into practice. In conversations with watermen in their boats and crab shanties and with women in the crab processing houses, I discovered that their most pressing concern was the threat to their existing way of life. I also discovered that watermen and women of faith believed that there is a scriptural foundation to steward the environment and its creatures, including the fish.

This provided a bridge for the community to understand and accept environmental stewardship ideas promoted by the regional environmental group, which they had considered secular and threatening. The community developed the Tangier Watermen's Stewardship Initiative, which included local government, school and church leaders and citizens.

After I spoke on biblical environmental stewardship and loving thy neighbor at a joint service of both local churches, 58 watermen bowed down at the altar and wept and asked God to forgive them for breaking the fishery laws and not being obedient to God. They then committed to the "Watermen's Stewardship Covenant." Many people's behavior and attitudes toward environmentalists, creation and the future changed radically and positively. Watermen even in their 70's and 80's—men who are not prone to change-rather than dumping trash overboard, brought it onto the island in bags. Government officials, scientists and environmentalists have been stunned by the dramatic change.

While not always agreeing with each other, some watermen and women are working with the Chesapeake Bay Foundation to restore the oyster reefs and to expand and diversify the local economy by developing oyster farming.

Used by permission of Susan Drake Emmerich. Susan Drake Emmerich, a member of the Au Sable Institute Board of Trustees, helped institute students to experience Tangier Island where she taught courses there in 1998 and 1999. Between Heaven & Earth: The Plight of the Chesapeake Watermen is a video about Ms. Emmerich's work on Tangier Island. It is available for purchase from the Au Sable Institute (www.ausable.org).

Introduction to Environmental Ethics

by Michael P. Nelson

Introduction

Ethics play an important role in our conversations and decisions about biodiversity. But what exactly are ethics? What does it mean to assert that there is an ethical dimension to a debate, or that ethics should inform our decision-making and our actions? This essay provides a basic framework for understanding ethics and a summary of the major ethical viewpoints on the environment and biodiversity.

What Are Ethics?

People tend to use the term *ethics* in two different ways.

Ethics help us decide how we ought to live. In their most general form, we might say that ethics are the standards we employ (among other factors) to determine our actions. They are *prescriptive* in that they tell us what we should or ought to do and which values we should or ought to hold. They also help us evaluate whether something is good or bad, right or wrong.

Ethics explain why things are important to us. Ethics are also concerned with how and why we value certain things and what actions properly reflect those values. In this sense, ethics appear more *descriptive*. Just as it is possible for taste to be a neutral and descriptive term—appreciation for a work of art can be a matter of *taste*—ethics can operate the same way. Hence, even though they clearly value nature differently, and therefore possess different environmental ethics, James Watt can be said to have an environmental ethic just as Aldo Leopold had one.

Either way, our ethics are not solely individual or deterministic: they are social constructs. This means that, while the raw capacity to extend moral consideration might be a product of our biology, our actual ethical beliefs are largely shaped by a cultural context and history.[1]

Grounding Ethical Claims

When someone offers an argument for or against protecting a threatened forest or river or plant, chances are that much of his or her argument will sound familiar. Places, species names, economic projections, etc., may all vary, but the logic of the argument will be similar, whether the debate is about beachfront condominium development in South Carolina or natural gas exploration in Wyoming. At a basic level, most of our arguments appeal to ethics: what is the right or wrong thing to do, what type of value do things hold, and why?

In turn, our ethical arguments—including those used in biodiversity debates—are often based on one of a number of established ethical theories. For instance, when someone argues that jobs are more important than environmental protection, he or

KEY POINTS

Ethics offer rules of conduct and ways of assigning value to and assessing the "rightness" of actions and things.

Most biodiversity debates reflect long-standing ethical assumptions and theories. Understanding these ethical grounds for arguments can help biodiversity advocates respond to the ethical underpinnings that inform most people's views on the environment.

The field of environmental ethics developed in response to the unique ethical problems presented by biodiversity loss, pollution, and other environmental issues. The major ethical theories in the field tend to distinguish themselves by the value they assign to nature and the actions they prescribe to address environmental problems.

she may be appealing to an ethical structure that goes back to the eighteenth-century English philosopher Jeremy Bentham and his theory of Utilitarianism (see below). Understanding the roots of these theories helps us to understand where people are coming from when they say that an action is good or bad, right or wrong. It also helps us understand how to counter their perspective, if necessary. What follows are thumbnail sketches of some of the leading Western ethical theories that continue to shape and define people's views on the environment today.

Utilitarianism
In its most basic form, utilitarianism suggests that we ought to judge an action, or decide upon a course of action, on the basis of the utility, happiness, or pleasure that action produces. The phrase "the greatest good for the greatest number" is often associated with this theory. For example, a developer who justifies a new sprawling residential development on the basis that it will provide housing for many families or increase the local tax base is appealing to a utilitarian theory.

Responses: Utilitarian justifications must often confront certain problems. First are problems of measurement: How do we assign values to the possible outcomes of our actions? Do all pleasures count equally? Is all happiness identical? Second, are problems of consequence: How do we know what the consequences of our actions might be? Should we justify horrific practices—such as slavery, child labor, or the destruction of the Amazon rainforest—because such practices might be useful or produce the most overall utility?

Rights Theory
Often seen as a reaction to utilitarianism, these ethical theories claim that we should adhere to certain rules or guiding principles that define an action as good or right when determining whether that action is right or wrong, irrespective of its consequences. For example, if someone argues that people should not be enslaved regardless of the benefits of slavery, they are basing their argument on the principle that people have a basic right to freedom that applies in all circumstances and overrides all consequences. Some have argued that the Endangered Species Act grants such basic rights (the right to continue to exist) to all species quite apart from their economic value.

Responses: Rights theorists must respond to two important questions. First, how do we sort out conflicts among or between principles or rights?[2] Second, how do we ultimately justify or establish those duties or rights that we decide are fundamental?

Divine Command/Natural Law
Divine command theory suggests that ethical precepts are the product of divine or revealed dictate (i.e., ethical rules are dictated from above by God or Krishna or Allah).[3] For example, we may believe that our stewardship of the land (or even our malicious impact upon it) is the morally correct course of action because it is what God intended; or we may believe it is morally right because humans are by nature stewards, caretakers, and nurturers and that the land is a proper object of this natural caretaking role. A variant of this theory—natural law—suggests that ethical precepts are a result of uncovering and then following the dictates of nature: in other words, that which is moral is often seen as that which is natural.

Responses: Adherents of these theories must be prepared to consider several questions. First, can we accept certain presuppositions in order to believe this theory? For instance, do we accept the existence of a divine being to give us instructions or a clear idea of that which is natural? Second, how do we know what is and what is not the will of God or what is natural? Who's to say, and how do we know we have it

Ethical Theories in Practice

Utilitarianism
"Our mission, as set forth by law, is to achieve quality land management under the sustainable multiple-use management concept to meet the diverse needs of people."
   USDA Forest Service Mission Statement (from the Forest Service Web site: www.fs.fed.us/fsjobs/forestservice /mission.html)
   The Forest Service's "Land of Many Uses" motto has often been interpreted along strictly utilitarian lines, emphasizing land management priorities that principally serve the economic and recreational needs of people.

Rights Theory
"To live free from harm, and the fear of harm, by human beings is the fundamental right of all sentient beings." Article One, the Universal Charter of the Rights of Other Species, by Lawrence Pope, the Charter Project/the Australian and New Zealand Federation of Animal Societies (www.melbourne.net/animals_australia /specials/charter.html).

Divine Law
"Almighty God envisioned a world of beauty and harmony, and he created it, making every part an expression of his freedom, wisdom, and love . . . If we examine carefully the social and environmental crisis which the world community is facing, we must conclude that we are still betraying the mandate God has given us: to be stewards called to collaborate with God in watching over creation in holiness and wisdom."
   From "Joint Declaration on Articulating a Code of Environmental Ethics," issued by Pope John Paul II and Ecumenical Patriarch Bartholomew of Constantinople, on June 10, 2002.

Natural Law
"Made from the Best Stuff on Earth' Snapple's array of tea and fruit beverages are made from all natural ingredients." —Cadbury Schweppes, Inc.
   At the supermarket you're likely to find hundreds of products that tout their "natural ingredients." This common marketing strategy is based on natural law theory: that actions and things that are derived from or found in "nature" are by definition superior to those that are not.

Virtue Theory
"Every person has a role to play in saving our planet. Action begins with a personal commitment. One person's commitment is the first step toward saving the planet for future generations, towards a living planet. You really can make a difference."
World Wildlife Fund —New Zealand Web site (www.wwf.org.nz)

Moral Sentiments
"Meat is Murder," Morrissey, former lead singer of the British rock band, The Smiths.

right? Third, how do we decide which messages or dictates, among many (some even contradictory) possible, are the correct ones to adhere to? Finally, does such an approach threaten to become less a matter of ethics, than one of merely following rules?

Virtue Theory
Some people hold to the belief that in general good people will perform good actions (as an extension of their goodness and perhaps as a way of attaining their own true happiness) and that they will help promote the well-being of all. Therefore, we need to maximize those qualities within people that make them virtuous. Although this appeal to ethics is not as popular as the others in environmental ethical discourse, it does occur. We conservationists often speak of nurturing the qualities or virtues of humility and respect within humans, and especially as humans interact with nature, with the assumption that by and large the humble and respectful person will act morally.

Responses: Clearly such a theory assumes a great deal: It assumes the ability of humans to foster various virtues; it assumes our ability to foster the correct ones; it assumes our strength of will to remain virtuous in tough spots; and it assumes that the actions of the virtuous person will in fact be environmentally ethical.

Moral Theory
This theory holds that we are ethical creatures because we are both rational and emotional creatures. If we reason that something commands our moral recognition, our moral sentiments (sentiments like compassion, sympathy, empathy) are prompted and spur our willingness to value that something and act on its behalf. Environmental philosopher J. Baird Callicott, for instance, has argued that it is such a theory of morality that underlies the Land Ethic of Aldo Leopold.[4] We see this theory when Leopold characterizes ethics as a product of both conscience and feeling:

> Obligations have no meaning without conscience, and the problem we face is the extension of the social conscience from people to land. No important change in ethics was ever accomplished without an internal change in our intellectual emphasis, loyalties, affections, and convictions.[5]

Responses: Theories of moral sentiments can be criticized for being overly subjective or relativistic and hard to pin down, for lacking prescriptive force since they seem at first glance only to describe the moral system at work, and for reducing ethics to a matter of biological determinism.[6]

Environmental Ethics

Environmental ethics is a new area of study within the larger and older field of ethics. In the early 1970s, a small cadre of philosophers began to realize that underlying our concern for and discussions about land use, biodiversity loss, and pollution were very real, interesting, and new ethical questions. We also began to see that complex philosophical notions lay at the core of our disagreements about what we should do with land, how we should value other species, and which policies we should enact to mitigate pollution. We quickly realized that environmental issues are

* *Environmental issues are inherently and intractably philosophical and ethical issues.*

inherently and intractably philosophical and ethical issues.

Those outside of philosophy soon recognized how critical the work of environmental ethics and environmental ethicists was to natural resource issues. Courses in environmental ethics were promptly required for natural resource majors in college, and environmental ethicists were granted joint appointments in humanities and natural resources departments; we were asked to sit on natural resource advisory boards and editorial boards of natural resource journals, invited to participate in and join scientifically oriented organizations and conferences, and asked to contribute articles to journals and chapters to textbooks in conservation biology, forestry, and other natural resource areas.

As the subdiscipline has evolved over the past three decades, environmental philosophers have separated into a number of distinct camps. Such camps distinguish themselves most profoundly by the value that they assume nature possesses and hence by the method or standard by which they believe we ought to go about addressing our environmental woes.

Anthropocentrism

Anthropocentrists are those who believe that environmental policies ought to be motivated and justified by their effect upon humans.[7] Of course, these philosophers often recognize both the full range of human values and the fact that human well-being is intimately entwined with the well-being of at least certain parts of the nonhuman world. For them, the nonhuman world is valuable only insofar as it affects humans. For the anthropocentrist, only humans possess intrinsic value; all else is valuable only for its utility for people. Anthropocentrists, then, agree with Immanuel Kant, who argues that "all duties towards animals, towards immaterial beings and towards inanimate objects are aimed indirectly at our duties towards mankind," or John Passmore, who claims that, "the supposition that anything but a human being has 'rights' is…quite untenable." For the anthropocentrist, we ought to be concerned about the loss of biodiversity and act to mitigate it only because such loss does or might negatively affect human beings. Plant biodiversity in the rainforest is valuable, they might argue, because it might provide cures for certain human diseases.

Zoocentrism

Zoocentrists are environmental philosophers who believe that, in addition to humans, certain nonhuman animals possess intrinsic value and garner direct moral standing.[8] These animal-centered, or zoocentric, ethicists argue that for all the reasons that we consider humans as intrinsically valuable, logical consistency dictates that we ought also to value certain nonhumans as intrinsically valuable, given only that these nonhuman animals possess the same trait that makes humans morally relevant. For the zoocentrist, humans and certain nonhuman animals possess intrinsic value; all else maintains only instrumental value. Hence, the zoocentrist is concerned about the loss of biodiversity because of the actual and potential negative impact that it has on both humans and certain nonhuman animals. For instance, they would hold that rainforest biodiversity preservation is important because it might provide cures for diseases in both human and certain nonhuman animals.

Biocentrism

Some philosophers have argued that the only way to avoid logical moral inconsistency is to include within the moral community all individual living things.[9] These life-centered, or biocentric, thinkers argue for the direct moral standing and intrinsic value of all individual living things, leaving only nonindividual living things as possessive of merely instrumental value. Albert Schweitzer, perhaps the most popularly recognized biocentrist, summarizes the position thus:

Ethics thus consists in this, that I experience the necessity of practicing the same reverence for life toward all with a will-to-live, as toward my own. Therein I have already the needed fundamental principle of morality. It is good to maintain and cherish life; it is *evil* to destroy and check life.[10]

Hence for the biocentrist, concern for, or policy regarding, biodiversity degradation is motivated and justified by the impact that it might have on *all* individual living things: we ought to be concerned about biodiversity loss because of the effect it has on humans, fish, and trees.

Biocentrism has been associated with Deep Ecology—a popular philosophy that sees humanity as a part of nature, rather than apart from or superior to it. Deep Ecology is also related to Universalism and Ethical Holism (discussed below).

Universalism
Some philosophers have gone so far as to argue that the only sensible and logically consistent moral community would be inclusive of all individual things, whether living or not. Those advocating this "universal consideration" suggest that we live in a morally rich world where everything is imbued with intrinsic value and direct moral standing. As Thomas Birch argues,

> Universal consideration—giving attention to others of all sorts, with the goal of ascertaining what, if any, direct ethical obligations arise from relating with them—should be adopted as one of the central constitutive principles of practical reasonableness.[11]

For these philosophers, our reaction to biodiversity loss or policy proposals attempting to curb it ought to be motivated not only by the impact that such loss has on all living things, but also by the impact that such loss has on even nonliving things such as mountains or rocks.

Ethical Holism
Some philosophers, including deep ecologists, have reacted against the atomism or individualism of all the above approaches. They have argued that the biosphere as a whole, as well as the systems that constitute it, deserve moral consideration, based on holistic understandings of natural systems derived from the science of ecology.[12] Although their approaches and arguments vary, this ethical holism refocuses moral concern on maintenance of the health of biotic communities, species, ecosystems, and even the Earth as a whole (if one were to extend this idea as far as James Lovelock's Gaia Hypothesis). Aldo Leopold expresses the most recognized version of ethical holism when he asserts that, "A thing is right when it tends to preserve the integrity, stability, and beauty of the biotic community. It is wrong when it tends otherwise."[13] Thus, biodiversity loss is a matter of concern because the health of species as well as specimens, watersheds as well as rivers, and forest ecosystems as well as individual trees is negatively affected.[14]

Ecofeminism
These philosophical discussions have spawned a variety of interesting and exciting areas of specialty. For example, "ecofeminism," as defined by leading ecofeminist scholar Val Plumwood, "is essentially a response to a set of key problems thrown up by the two great social currents of the later part of this century—feminism and the environmental movement—and addresses a number of shared problems."[15] Ecofeminists have developed insightful analogies between the historical oppression of nature by humans, and that of women by men. They have suggested that Western environmental problems should be, perhaps even that they can only be, understood in light of a larger historical attempt to bifurcate the world in such a way that women and nature are linked with what is morally degraded or downgraded, and that men and the nonnatural are conceptually tied to the morally relevant or superior.

Environmental Justice

Other thinkers have focused on how various forms of environmental degradation, and even various proposals to remedy this degradation, play out in terms of justice between and within societies. Critiquing such concepts as Gross National Product (GNP) as a measure of progress, capitalism and free market economics, technological fixes to environmental problems, the imposition of wilderness areas and parks on local populaces, and economic development, those interested in issues of environmental justice (or ecojustice) have dramatically illustrated the negative global result of our current environmental problems, and especially how the costs of environmentally negligent behavior are unfairly borne by some. As philosopher Peter Wenz puts it, "questions about justice arise concerning those things that are, or are perceived to be, in short supply relative to the demand for them." Given that the Earth's resources are finite, and given that we are all concerned with getting our fair share of those resources, environmental issues and ethics are inherently a matter of justice.[16]

Notes

[1] One must be careful not to assume that this means that ethics are relativistic; that conclusion does not necessarily follow from the premise that ethics are social constructs. The value of the dollar is also a social construct, but it is very real and very objective nonetheless.

[2] On this see, "Rights and Responsibilities: What Obligations Do We Owe to the Natural World (and Each Other)," at page [#].

[3] Some might suggest that raw appeals to the "laws of nature" might also fit within this category.

[4] See especially J. Baird Callicott's work in *In Defense of the Land Ethic and Beyond the Land Ethic* (Albany, NY: State University of New York Press, 1989 and 1999, respectively).

[5] Aldo Leopold, *A Sand County Almanac: With Essays on Conservation from Round River* (New York, NY: Ballantine Books, 1966), 246.

[6] Many of these concerns as they apply to the land ethic have been addressed by Callicott in his essay, "Can a Theory of Moral Sentiments Support a Genuinely Normative Environmental Ethic?" in *Beyond the Land Ethic: More Essays in Environmental Philosophy* (Albany, NY: State University of New York Press, 1999), 99-115.

[7] See William Baxter, *People or Penguins: The Case for Optimal Pollution* (New York: Columbia University Press, 1974); John Passmore, *Man's Responsibility for Nature*, 2nd ed. (London, UK: Duckworth, 1980); and Bryan Norton, *Why Preserve Natural Variety?* (Princeton, NJ: Princeton University Press, 1988).

[8] See Peter Singer, *Animal Liberation*, 2nd ed. (New York, NY: Random House, 1990); and Tom Regan, *The Case for Animal Rights* (Berkeley, CA: University of California Press, 1983).

[9] See Kenneth Goodpaster, "On Being Morally Considerable," *Journal of Philosophy* 75 (1978): 308-25; Robin Attfield, *The Ethics of Environmental Concern*, 2nd ed. (Athens, GA: University of Georgia Press, 1991); and Paul W. Taylor, *Respect for Nature* (Princeton, NJ: Princeton University Press, 1986).

[10] See Albert Schweitzer, *Civilization and Ethics, Part II, Philosophy of Civilization*, trans. John Naish (London, UK: A & C Black, 1923), 254.

[11] See Thomas Birch, "Moral Considerability and Universal Consideration," *Environmental Ethics* 15, no. 4 (Winter 1993): 313.

[12] See the works of J. Baird Callicott, mentioned above, and Arne Naess, *Ecology, Community and Lifestyle* (Cambridge, UK: Cambridge University Press, 1989); George Sessions and Bill Devall, *Deep Ecology: Living as if Nature Mattered* (Layton, UT: Gibbs-Smith, 1985); and Warwick Fox, *Toward a Transpersonal Ecology*, 2nd . (Albany, NY: State University of New York Press, 1995).

[13] Aldo Leopold, *A Sand County Almanac: With Essays on Conservation from Round River* (New York, NY: Ballantine Books, 1966).

[14] In fact, for the ecocentrists, biodiversity itself is more than just a collection of various individual living things; they construe it in a far more holistic fashion than any of the other more individualistic environmental ethical theorists do.

[15] See Val Plumwood, *Feminism and the Mastery of Nature* (London, UK: Routledge, 1993); and Karen Warren, "The Power and the Promise of Ecological Feminism," *Environmental Ethics* 12, no. 2 (1990): 125-46.

[16] See Peter Wenz, *Environmental Justice* (Albany, NY: State University of New York Press, 1988); Ramachandra Guha, "Radical American Environmentalism and Wilderness Preservation: A Third World Critique," *Environmental Ethics* 11, no. 1 (1989): 71-83; and Vandana Shiva, *Staying Alive: Women, Ecology, and Development* (London, UK: Zed Books, 1989).

2,000 Years of Western Ideas About Nature in Less than 2,000 Words[1]

by Michael Nelson

People do not do or believe things "just because." Our ethical sensibilities, our ideas, and our assumptions about reality flow from the past. This chart is an attempt to present a rough overview of the sources of the conceptual and ethical ideas embodied in Western culture. In all fairness, the reader should be forewarned that intellectual history is a complex web that, by its very nature, resists this sort of reduction. Hence, any such presentation will, of necessity, be incomplete. The point is, however, to see and be able to make sense of the origins of our ideas about humans, nature, and what constitutes an appropriate human/nature relationship. Such awareness provides us with an understanding of one another, an important step toward working together.

Ethical/Metaphysical/Spiritual Belief	Origin of Belief	What Does it Mean?	Implications
Nature is messy and inefficient.	Ancient Greeks (6th century BC), David Hume (1711-1776)	Purity of form is an expression of that which is good.	Nature ought to be "neatened up" or ordered by humans
Nature is knowable and quantifiable; humans can control it (manifests itself in Atomism, Materialism, and Mechanism).	Ancient Greeks (6th century BC), John Locke (1632-1704), Francis Bacon (1561-1626), certain views of science	We know nature by taking it apart, by knowing it we gain control of it, by gaining control of it we increase our mastery of it.	We should learn about nature in order to control it; the whole of nature is nothing but the sum of its parts—no more, no less.
Dualism	Pythagoras (580-500 BC) who influenced Socrates (470-399 BC) and Plato (427-347 BC), Christianity, René Descartes (1596-1650), most contemporary Westerners.	The mind and body are distinct, at least for humans (sometimes seen as a reaction to Atomism, as a way to save human uniqueness).	Humans are separate from and special with regard to nature. The real nature of humans is otherworldly; all else is of this world and the bodily realm.
Nature is suspect, dangerous, the realm of Satan.	Puritanism, Jonathan Edwards (1703-1758), Cotton Mather (1584-1652))	Untamed nature is Satan's foothold, is not only without value, but is of disvalue, is bad, even evil; those humans associated with the world of nature (Pagans, American Indians) are also bad.	The role of humans is to rid the world of room for the devil as well as everything associated with this realm, transform the "natural" into the "artificial," prepare the world for God, fulfill its proper mission.

ORIGINS AND ROOTS: A CRASH COURSE IN THEOLOGICAL AND ETHICAL PERSPECTIVES ON BIODIVERSITY

Application to Biodiversity	Quotation
The results of the straightened streams, monoculture pine plantations, and filled-in wetlands generally have a negative impact on biodiversity.	"Nature hates calculators." —Ralph Waldo Emerson
Although it's important to learn about nature/biodiversity, it's important only insofar as it helps us control it.	"Knowing the nature and behavior of fire, water, air, stars, the heavens, and all the other bodies which surround us…we can employ these entities for all the purposes for which they are suited, and so make ourselves masters and possessors of nature." —René Descartes
Sometimes employed as a justification for anthropocentrism, it explains and sanctions our indifference to the biologically diverse world around us.	"One can no more ask if the body and the soul are one than if the wax and the impression it receives are one." —Aristotle "It is certain that I (that is, my mind, by which I am what I am) is entirely and truly distinct from my body, and may exist without it." —René Descartes
Biodiversity is clearly associated with the "natural"; either it's therefore bad, or it's of little importance.	"Nature red in tooth and claw" —Tennyson "Nature is a hanging judge." —Anonymous

ORIGINS AND ROOTS: A CRASH COURSE IN THEOLOGICAL AND ETHICAL PERSPECTIVES ON BIODIVERSITY

Ethical/Metaphysical/ Spiritual Belief	Origin of Belief	What Does it Mean?	Implications
Dominion	Genesis, Francis Bacon (1561-1626); later becomes cornerstone of John Calvin (1509-1564) and Calvinism.	Humans are to have dominion over nature; the human/nature relationship is despotic. This results in the idea that we work to produce and to consume in order to confirm our salvation.	Humans are not only separate and special, but are commanded to "dominate and subdue" nature, to turn the natural into the civilized. We produce and consume; therefore we are good.
Nature and humans are a miracle called Creation.	Genesis, John Muir (1838-1914)	We are all meant to be here and we are valued by God.	Both humans and nature are spiritually significant and possess value; both are the handiwork of God
Stewardship	Reinterpretation of Genesis I	Humans are properly viewed as stewards or caretakers of nature.	The human/nature relationship ought to be a caretaker relationship, with humans in charge but for the good of the Earth.
Natural Law/Divine Command Theory	Aristotle (384-322 BC), early thinkers in various religious traditions	Right and wrong is dictated or determined by some higher entity, either nature itself or some divine being.	That which is natural is that which is right or good, or that which God commands is that which is good; both can be discovered.
Rights Theory (one of the dominant contemporary ethical theories)	Immanuel Kant (1724-1804)	Humans have certain fundamental rights (e.g., to continue to exist, to be free, etc.); all else is only a means to secure human ends.	Humans are valuable and distinct as ends in and of themselves; nature is valuable only insofar as it provides for the fulfillment of basic human rights (whatever they might be).
Utilitarianism, which manifests itself as Capitalism and Neo-Classical economics (one of the dominant contemporary ethical theories).	Jeremy Bentham (1748-1832), John Stuart Mill (1806-1873), Adam Smith (1723-1790)	Nature is valuable only insofar as it provides for the utility or happiness or well-being of human beings. The market determines what is of value or has utility.	Nature has value insofar as it secures the well-being of human ends. Decision making is driven only by market value—if it can't pay for itself, it can't be saved.

Application to Biodiversity	Quotation
Biodiversity loss might be bad if it negatively affects humans, but it might also be seen as good because it's an indication that humans are successfully dominating and subduing nature. Idle land is bad land (John Muir's father, Daniel Muir—a Calvinist minister—believed this).	"Let us make man in our own image…and let them be masters of the fish of the sea, the birds of the heaven, the cattle," etc. "Be fruitful and multiply, fill the earth and conquer it. Be masters of the fish of the sea", etc. —Genesis I "Nature's secrets must be tortured out of her." —Francis Bacon
It is wrong to undo God's handiwork by facilitating the loss of biodiversity.	"And God said, 'Let the earth bring forth vegetation, plants yielding seed, and fruit trees bearing fruit in which is their seed, each according to its kind,'…'Let the waters bring forth swarms of living creatures, and let the birds fly above the earth'…'Let the earth bring forth living creatures according to their kind'…And it was so…And God saw that it was good." —Genesis "God himself seems to be always doing his best here, working like a man in a glow of enthusiasm." —John Muir
Humans have a moral obligation to tend to the preservation of biodiversity as a function of their role as stewards.	"The Lord God took the man and put him in the garden of Eden to till it and keep it." —Genesis
Biodiversity is important or good if we conclude that it is natural or if God says it is; biodiversity loss is a matter of indifference if we conclude that it is natural or if God says it is unimportant.	"If one way be better than another, that you may be sure is Nature's way." —Aristotle "Never does Nature say one thing and Wisdom another." —Juvenal "Deviation from Nature is deviation from happiness." —Samuel Johnson
Biodiversity is valuable only insofar as it provides for the fulfillment of certain human rights; it is indirectly valuable at best.	An action is right if it "signifies consistency with the will of God." —William Paley "Act only on that maxim whereby thou canst at the same time will that it should become universal law." —Immanuel Kant "Safeguarding the rights of others is the most noble and beautiful end of a human being." —Kahlil Gibran
Biodiversity is valuable only insofar as it secures or provides for the well-being of humans, and is economically profitable. Protecting biodiversity is purely and solely a matter of cost-benefit analysis.	"The principle of utility approves or disapproves of every action, whereby it tends to produce benefit, advantage, pleasure, good, or happiness or to prevent the happening of mischief, pain, evil, or unhappiness to the party whose interest is considered." — Jeremy Bentham "Nature is not a temple but a workshop in which man is the laborer." —Ivan Turgenev

ORIGINS AND ROOTS: A CRASH COURSE IN THEOLOGICAL AND ETHICAL PERSPECTIVES ON BIODIVERSITY

ORIGINS AND ROOTS: A CRASH COURSE IN THEOLOGICAL AND ETHICAL PERSPECTIVES ON BIODIVERSITY

Ethical/Metaphysical/ Spiritual Belief	Origin of Belief	What Does it Mean?	Implications
Wise-Use Conservation (not to be confused with the current Wise-Use movement)	Adaptation of utilitarianism, Gifford Pinchot (1865-1946)	Nature provides for human well-being.	Nature is valuable only insofar as it provides for human well-being, but should be managed to maintain a reliable supply of natural goods.
Cornucopia	Julian Simon (1932-1998)	Nature is unlimited in its ability to provide resources and absorb impact.	Humans can do whatever they please with regard to nature since nature is ultimately forgiving.
Holism	Evolutionary theory (Charles Darwin [1809-1882]), ecology, new physics quantum theory, relativity theory (Albert Einstein [1879-1955]) interpreted by Aldo Leopold (1887-1948), J. Baird Callicott, Fritjof Capra, Deep Ecologists, Paul Shepard (contemporary)	The whole is more than the sum of its parts; relationships among parts also have a fundamental reality, ecological wholes (species, ecosystems, biotic communities, watersheds) exist and have moral value.	The world, parts of the world, and parts of those parts, are systemically related and integrated with one another; thus ethical systems cannot merely account for individuals; human well-being is provided for by tending to the well-being of the systems of which they are a part and upon which they depend.
Gaia Hypothesis	James Lovelock, Lynn Margulis (contemporary)	The earth itself is, or is like, a living organism in that it has the ability to sustain itself amidst external change, absorb and mitigate impact, and secure its own health.	This challenges our concept of "individual living thing"; ecological systems are crucial to the functioning of the larger Earth system; parts of the whole are ultimately important.
Ecofeminism	Val Plumwood, Karen Warren, Rosemary R. Ruether (contemporary)	A strong parallel exists between the historical oppression of nature by humans and historical oppression of women by men; instances of oppression are manifestations of a similar conceptualization and the logic of domination.	To the extent that environmental problems are problems of humans dominating nature, they are conceptually linked to other oppressive structures. The focus should be on oppressive conceptual structures in general, not merely particular representations of them.

Application to Biodiversity	Quotation
Biodiversity is good as a source of natural resources, medicine, recreation, etc.	"The use of natural resources for the greatest good of the greatest number for the longest time." "There are just two things on this material earth—people and natural resources." —Gifford Pinchot
There seems little need to be worried about the human impact on biodiversity.	"Constraints are set by political and economic, not ecological or physical, facts." —William W. Murdoch "The major constraint upon the human capacity to enjoy unlimited minerals, energy, and other raw materials…is knowledge. And the source of knowledge is the human mind…this is why an increase of human beings…constitutes a crucial addition to the stock of natural resources." —Julian Simon
Biodiversity is seen as more than merely collections of individual specimens or a variety of species types; it is far more inclusive, thus biodiversity preservation is important, even intrinsically valuable.	"What makes it so hard to organize the environment sensibly is that everything we touch is hooked up to everything else." —Isaac Asimov "A thing is right when it tends to preserve the integrity, stability, and beauty of the biotic community. It is wrong when it tends otherwise." —Aldo Leopold
Preservation of biodiversity is ultimately important, for t provides for the mechanism for global health, which secures the health of everything that is a part of Gaia.	"[The Gaia Hypothesis holds that] the nonliving and the living represent a self-regulating system that keeps itself in a constant state." —James Lovelock "Earth is a single huge organism intentionally creating an optimum environment for itself." —Richard Kerr
Loss of biodiversity is but one example and manifestation of an inappropriate conceptualization of a relationship (the human/nature relationship in this case). Preservation of biodiversity cannot be seen in isolation from other oppressive systems.	"We cannot criticize the hierarchy of male over female without ultimately criticizing and overcoming the hierarchy of human over nature" —Rosemary R. Ruether "Does the wanton subjugation of nature by our species have a causal connection with the wanton subjugation of women by men?" —David Quammen

[1] See Joy A. Palmer, ed., *Fifty Key Thinkers on the Environment* (London, UK: Routledge, 2001); this is an excellent reference resource for the ideas and publications of many who have historically shaped environmental thought.

SECTION III

How Shall We Live? Applying Ethical and Religious Perspectives to the Biodiversity Crisis

The Ways We Value Nature

by Michael P. Nelson

Introduction

In environmental debates, questions of value are sometimes the deciding factor upon which an issue is decided. But how do we ascribe value to a flycatcher or an oak savanna? For most biodiversity advocates, the natural world is rich in value and importance. But the practical conservationist understands that not all value is created equal, and not everyone values nature in the same ways. The more clearly we can articulate and defend nature's value, both in terms of what it does for us and its value for its own sake, the more successful we will be in protecting and restoring it.

We tend to value nature in three ways: utilitarian value, intrinsic value, and non-substitutable value.

Utilitarian Value

First, and most obvious, nature possesses what is often referred to as *Utilitarian Value*. Most generally stated, this means that nature has value because it serves as a means to achieve some other end.

The debate over oil exploration in the Arctic National Wildlife Refuge is an example of the way we understand nature based on utilitarian value. For drilling proponents—the oil industry, the state of Alaska, and its congressional delegation, among others—the Refuge is defined solely by its usefulness: the potential oil it may produce and the alleged economic and national security benefits that would flow from it. Yet drilling opponents have also cited utilitarian values in their opposition to oil exploration. The Gwich'in nation still depends on caribou for subsistence and has opposed development of the Refuge because of the impact it would have on its people's ability to hunt food. Other opponents have cited the critical habitat the refuge provides and the challenging outdoor recreation it offers as reasons to protect it.

Utilitarian value is a kind of value that is wholly substitutable, meaning that it can potentially be met in other ways. For example, proponents of oil drilling in the Refuge have claimed that caribou and polar bear will move their calving and denning areas away from drilling rigs, pipelines, and waste sites; some have suggested that in the 21st century, no people need to depend on subsistence hunting for food. Therefore, because of the substitutability of utilitarian value, they argue, we cannot rest the case for something like wilderness preservation on purely utilitarian grounds.

However, if we are going to include utilitarian values in our assessment of an environmental decision, then we need to take into account the full range of useful values that a natural area or a species provides. For example, a forest provides more than two-by-fours; it offers a range of other goods and services, from water purification to erosion control to carbon-storage. If a developer wants to argue that the value of a forest tract should be reduced to its "bottom line" utilitarian value, then full-cost accounting of all the values the forest provides, including its recreational value and its value for wildlife habitat, is the only honest and acceptable way to proceed.

Intrinsic Value

The term *Intrinsic Value* is most commonly used to describe value inherent in something regardless of its usefulness or benefit.

KEY POINTS

There are three important types of value that people ascribe to the environment: utilitarian, intrinsic, and nonsubstitutable.

Utilitarian value is based on the idea that nature has value because it serves as a means to achieve some other end.

Intrinsic value recognizes the worth of something in itself, putting the burden of proof on anyone who would attempt to override that value to offer a compelling reason to do so.

Nonsubstitutable value takes into account the uniqueness of a place, thing, or experience being valued, but the value assigned to that unique characteristic is determined by the ends that the object can help achieve, rather than its intrinsic worth.

Our feelings for certain things clearly go beyond their utilitarian value. For example, few of us would say that our loved-ones are important based on their usefulness (their utilitarian value). Instead, they are valuable because they are good, or have value, in themselves. Likewise, most ethical and religious traditions foster a belief that all of humanity possess a certain kind of worth or dignity that transcends any utilitarian purpose they may serve.

Of course, things can possess more than one type of value (intrinsic value is value *in addition to*, not value *apart from*, other values). For example, we can recognize that monarch butterflies have an intrinsic right to exist, while also appreciating their beauty and the important role they play as pollinators.

Can we define nature, or biodiversity, as intrinsically valuable? Many ethicists and philosophers have argued that nature has intrinsic value.[1] The following summary arguments make this case:

- We (human beings) value ourselves intrinsically.
- We assume that we possess intrinsic value because we are living beings that have interests (conscious as well as nonconscious) that can be subverted or nurtured, we feel pain and pleasure, we are self-conscious, etc.
- In American history, we have not always extended intrinsic value to all people. But as our knowledge changed about the qualities of those whom we previously considered outside our moral realm (and not worthy of intrinsic value), we came to realize that for the same reasons that we believed that we had intrinsic value, we had to extend intrinsic value to others as well.
- An extension of intrinsic value to include all living things, then, seems inevitable. If we believe that all things that are alive have interests, and if we believe that all things with interests ought to be granted intrinsic value (since that is why we believe that we have intrinsic value), then we must believe that all living things have intrinsic value.
- As rational creatures, we are compelled by the force of logic to grant all living things intrinsic value.

Beyond granting intrinsic value to other

※ If a developer wants to argue that the value of a forest tract should be reduced to its "bottom line" utilitarian value, then full-cost accounting of all the values the forest provides, including its recreational value and its value for wildlife habitat, is the only honest and acceptable way to proceed.

How does the public value nature?

How does the American public value nature? On the Biodiversity Project's 2002 national biodiversity poll, respondents were asked to rate the importance of different values that underlie support for the environment (on a scale of 1-10, where 1 indicates "not at all important" and 10 indicates "very important"). While the poll did not test the appeal of utilitarian, nonsubstitutable, and intrinsic values per se, we did ask questions that could be roughly grouped within these categories.

In the poll, messages based on utilitarian value ("leave the earth in good shape" for people in the future and "protect nature for you and your family to enjoy a healthy life") and non-substitutable value ("appreciation of the beauty of nature") ranked higher than intrinsic value ("a respect for nature for its own sake").

Of those surveyed, 47% said that "a respect for nature for its own sake" was extremely important to them personally (compared to 56% who said that responsibility to future generations, the highest ranked value, was extremely important to them). Women, African-Americans, Hispanics, and 40- to 59-year-olds tended to rank "intrinsic value" higher than other demographic groups.

Values Underlying Support for Protecting the Environment
% saying "extremely important"

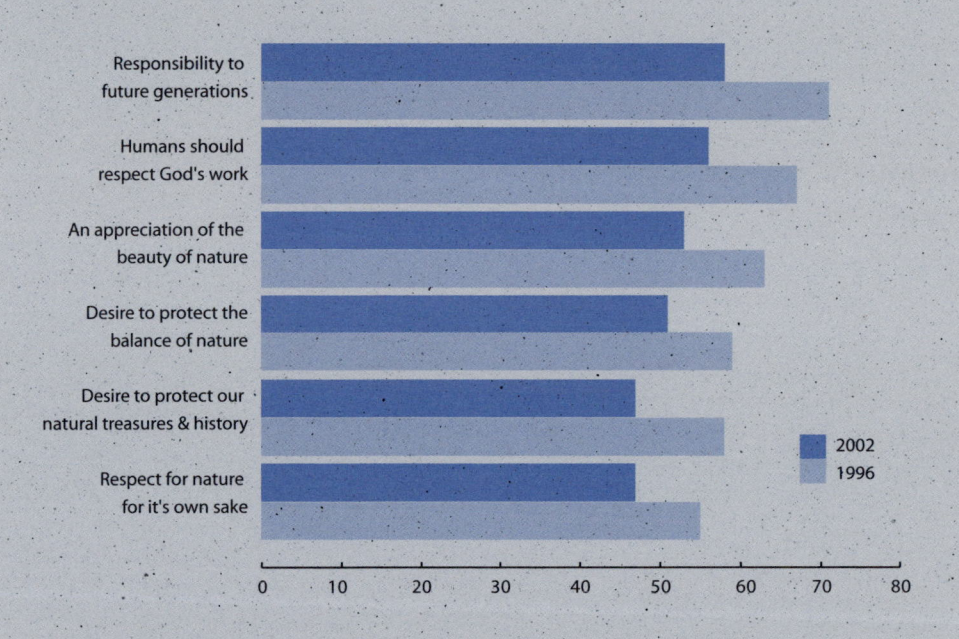

Value	2002	1996
Responsibility to future generations	~57	~71
Humans should respect God's work	~55	~66
An appreciation of the beauty of nature	~52	~62
Desire to protect the balance of nature	~50	~58
Desire to protect our natural treasures & history	~47	~58
Respect for nature for it's own sake	~47	~55

Q. Please think of a 1 to 10 scale. This time 1 means something that is *not at all* a reason to you personally and 10 means it is an *extremely important* reason to you personally to care about protecting the environment: Here's the first one: How important is this to you personally as a reason to care about protecting the environment?
a. Nature is God's creation and humans should respect God's work.
b. A respect for nature for its own sake.
c. A personal responsibility to leave the earth in good shape for future generations.
d. An appreciation of the beauty of nature.
e. A desire to protect the balance of nature for you and your family to enjoy a healthy life.
f. A desire, as an American, to protect our country's natural treasures and natural history.
Source: Beldon, Russonello and Stewart, *Americans and Biodiversity: New Perspectives in 2002*, (The Biodiversity Project: Madison, WI, 2002), 14.

species, this argument has been extended to claim intrinsic value to ecosystems as well. The argument proceeds along the following lines. Recently ecology has shown us that we are parts of larger systems that themselves also possess those qualities that we have come to label as living. Therefore, it is rational to view these larger living systems (ecosystems, biotic communities, watersheds, etc.) as valuable in and of themselves or as intrinsically valuable. Some have even claimed that the Earth itself, which can be viewed as a self-regulating organism, is entitled to be considered as having intrinsic value.

What are the advantages of ascribing intrinsic value to nature? When something possesses intrinsic value, it becomes worthy of moral consideration *in its own right*—valuable in itself—not simply in relation to its benefit to something else. In other words, intrinsic value endows a thing with direct moral standing. Hence, a yellow-bellied flycatcher, a marsh marigold, and an alder swamp—as possessors of intrinsic value—can be said to occupy the same general moral space as a human being. Granting intrinsic values to nonhuman living things *shifts the burden* of proof on to those who would despoil the natural world and away from those who wish to preserve, protect, and defend it.

Of course the possession of intrinsic value is not regarded as an *absolute* moral trump. Intrinsic value can be legitimately overridden when there are compelling reasons. However, overriding someone's or something's intrinsic value is usually seen as a difficult task (the concern that granting intrinsic value to the natural world may compromise human interests is perhaps what drives some people to conclude that the nonhuman world cannot possess intrinsic value).

Non-Substitutable (Constitutive) Value

Utilitarian and intrinsic values are not the only ways people value things—there is another way, called constitutive (or in this essay) nonsubstitutable value. Something has nonsubstitutable value if it is a necessary and irreplaceable component of attaining some other desirable end.

To this extent, nonsubstitutable value is similar to utilitarian value in that an object has value because it is a means to an end. But nonsubstitutable value is different from utilitarian value because the object is valued due to its unique capacity to deliver that desired end. At the same time, however, the object is not valued intrinsically, or for its own sake. The object's value remains contingent upon the importance (perhaps usefulness) of the end that the person who is assigning value desires.

For example, in the debates over oil exploration in Alaska's Arctic Refuge discussed above, the Gwich'in nation opposes the destruction of the Refuge because its people see the land as inextricable from

* Non-substitutable value is important to keep in mind because it offers a way of capturing the powerful and complex emotional responses that people have toward nature that cannot be reduced to simple material usefulness.

their identity as a people. While other lands might meet their material needs, no other place can serve as their spiritual and cultural home. In a similar manner, drilling in the Refuge has been opposed by a majority of Americans, even though a tiny fraction of the public will ever benefit directly from the Refuge by visiting it. However, for many of those who have expressed opposition, protecting the refuge is important because it embodies an ideal of undisturbed wilderness, natural beauty, or the last American frontier, which is important to them and cannot simply be replaced by something else.

Nonsubstitutable value is important to keep in mind because it offers a way of capturing the powerful and complex emotional responses that people have toward nature that cannot be reduced to simple material usefulness. At the same time, nonsubstitutable value is "in the eye of the beholder." If people decided, for example, that the psychological importance of knowing that America still had wild places like the Arctic Refuge was no longer very important to them, then the Refuge would lose its value as a means to that end. Does this mean that the public as a whole is not receptive to arguments based on intrinsic value? No. First, even though intrinsic value ranked behind other human-centered messages, in the 1996 and 2002 Biodiversity Polls nearly a majority of Americans acknowledged nature's intrinsic right to exist. Second, as with any message, the appeal of an argument based on intrinsic value depends on to whom you are talking. But in the same way that intrinsic values are *in addition* to other values, an argument that invokes utilitarian, nonsubstitutable, and intrinsic values is likely to be more effective with more people than one that depends solely on intrinsic value.

Notes

[1] Some of the leading thinkers have included Warwick Fox, "What Does the Recognition of Intrinsic Value Entail?" *The Trumpeter* 10, no. 3 (1993); Freya Mathews, *The Ecological Self* (London, UK: Routledge, 1991), and Lawrence Johnson, *A Morally Deep World: An Essay on Moral Significance and Environmental Ethics* (Cambridge, UK: Cambridge University Press, 1991).

COMMUNICATIONS TIP

Here's an example of a message that appeals to the utilitarian, intrinsic, and nonsubstitutable values of nature:

"The wetlands outside our town are valuable to us because they filter our water and control the spring floods. A lot of us enjoy them for canoeing, fishing, hunting, and bird watching (UTILITARIAN VALUE). These wetlands are a unique and special place in their own right—the hundreds of species of plants and animals that live there have as much a right to a home as we do (INTRINSIC VALUE). But the wetlands are also an irreplaceable part of our community. How many of us remember catching our first frogs there, or exploring it as kids? Can the mini-mall they want to build there replace the beauty and mystery of our wetlands (NONSUBSTITUTABLE VALUE)?"

Rights and Responsibilities: What Obligations Do We Owe to the Natural World (and Each Other)?

by Michael Nelson and Robb Cowie

Introduction

The idea of rights figures powerfully in our society, shaping our laws, our political vocabulary, and our actions. Over and over, advocates for species and habitat protection must contend with someone's assertion that his or her rights take precedence over preserving habitat. To address these claims, we need to understand the premises that support them. By doing so, we can also make the case that people have obligations to act ethically toward the natural world.

Rights and Obligations

Historically, rights (to freedom, property, expression, etc.) are thought to come from a sacred source (e.g., conferred by God), or a secular authority (e.g., defined by collective human agreement and then written into law). Either way, rights often serve as moral trumps. That is, in the absence of a strong justification for limiting or overriding a right, that right is presumed to take precedence. (Of course, rights claims are not absolute; they can be overridden in extreme circumstances: in the classic example, the right to free speech does not give someone the right to yell, "fire!" in a crowded cinema.)

But rights do not stand alone; they demand corresponding obligations from everyone else. One person cannot claim to possess a right if others do not possess an obligation to at least recognize and respect that claim. At the same time, traditional rights theories hold that when we assert our rights, we also acknowledge our moral obligation to ensure similar rights for others. Rights and obligations go hand in hand.

Leaving Nature Out of the Equation?

The prevailing theories of rights in our culture—including theories where rights are derived from both sacred and secular sources—tend to view each individual human being as an autonomous and independent entity who is seeking to maximize his or her rights and who is in competition with other people who are trying to maximize their rights, too. These rights theories also tend to regard people as fundamentally different and separate from nature.[1] They deny any inherent connection between humans and their environmental context and insist that inalienable rights apply only to those separate, independent, and autonomous humans.

These assumptions pose some dilemmas. First, we know that logically we cannot maximize for separate variables. In other words, the real world imposes physical and social limitations that will prevent some, if not all, people from fully exercising all of their rights and interests. For example, Americans generally claim that it is their right to own property and to make their home on it. At the same time, even before the pernicious environmental, economic, and social effects of sprawl began to emerge, most communities had established ordinances that prevented landowners from building whatever they wanted, wherever they chose.

Why? We know that what one landowner does on his or her land can have an impact on another's property. If a landowner fills a wetland, he might cause his downstream neighbor's property to flood, because he has diminished the land's capacity to absorb rainfall. Therefore, not only do we enjoy basic human rights, but

> * Rights often serve as moral trumps. That is, in the absence of a strong justification for limiting or overriding a right, that right is presumed to take precedence.

KEY POINTS

Concepts of rights can have a profound influence on biodiversity debates, but rights do not stand alone—they are indivisible from responsibilities.

Evolutionary and ecological theories have challenged the basis of rights theories by emphasizing the connections between individuals and species.

Our connections to the larger ecological community require a responsibility to the natural world.

the social contract that guarantees our individual rights also demands of us a reciprocal moral obligation to acknowledge and ensure those same rights for others.

At a deeper level, many environmentalists reject the very premises of rights theory. Instead of seeing people as wholly autonomous individuals, separate from each other and from the natural world, they instead embrace a more holistic vision of our place in the world that is grounded in evolutionary theory and ecology. Evolution tells us that humans emerged from other life forms and that our environment has defined our physical (and some evolutionary theorists have argued our social) make-up. From an ecological perspective, all the species within the living world are interdependent and related within the "web of life." Individual plants, animals, species, and ecosystems are defined by and exist within a matrix of relationships. They owe their identity to, and they cannot be wholly separated out from, the complex web of relations that link them to the other members of the community of life. From this perspective, the claims of traditional rights theories make little sense, because no individual or species is autonomous from its connections to the biosphere.

This holistic perspective is not based simply on science. Many of the world's religious and spiritual traditions reflect this view, including major strains of the Judeo-Christian tradition. How then, do we reconcile the apparent discontinuity between traditional notions of rights and the more holistic perspective contained in many scientific and spiritual worldviews?

ETHICS IN THE REAL WORLD

The Klamath Basin Controversy: A Case Study in the Limits of Rights

In the Klamath Basin, along the Oregon-California border, drought has exacerbated conflicting demands for water, putting farmers and ranchers, native tribes, the down-river fishing communities, and environmentalists in conflict. In April 2001, the federal government denied water diversions for irrigation to prevent water levels in the basin from further dropping and thereby jeopardizing endangered short-nosed and lost river sucker fish, coho salmon, and other wildlife in Klamath Lake and its surrounding rivers and wetlands. The decision enraged agriculture-dependent local communities, creating another "zero-sum" controversy where, in this case, the survival of sucker fish (described as "worthless" and "trash" by some) was portrayed as coming at the expense of farmers' water rights.[4]

The controversy reveals the limitations of our traditional notions of rights, whereby autonomous individuals all attempt to maximize their own rights and self-interest. In reality, natural limits (such as water scarcity) make it impossible for all of us to exercise our rights without constraint. While Klamath farmers objected to federal action to protect fish, they were not the only human communities that had suffered. Federal irrigation projects in the Klamath Basin had substantially altered the ecosystem, contributing to the decline of suckers and salmon and hurting the salmon-fishing-based economies of communities downstream. As a result, commercial salmon interests demanded federal action to protect spawning grounds for fish. In addition, federal policies that helped make the Basin suitable for large-scale agriculture (and opened the surrounding forests to logging and mining) came at the direct expense of the Klamath tribes, who, under a 1954 law, lost over 21 million acres of their 22 million acre reservation lands to federal control.[5]

Missing in much of the discussion of the Klamath controversy is any mention of responsibility. From an ethical perspective, we have an obligation to acknowledge that others share the same rights that we assert: in this case, the right to a healthy environment and economic survival for all the Klamath communities—not just farmers or fishermen—from the headwaters of the Klamath River to the Pacific. This example also illustrates the connection between human rights and the environment: for the Klamath tribes, the suppression of their human rights (their right to self-determination) was the first step in their dispossession of the land and its conversion to unsustainable uses.

Leopold's Land Ethic acknowledges the

Extending Our Sense of Responsibility

We cannot simply jettison the concept of rights or ignore the influence of rights theory on our legal, political, and economic systems and our culture. But there are ways we can argue that we should extend our moral community beyond the human realm to nature. One way to do this is by talking about responsibility.

Most of us recognize that, to the degree that our actions affect others, they can be judged on an ethical scale. But why is it that actions that affect others become subject to ethical judgment? It is because those others live with us in a social community and therefore an ethical community. That is, we grant moral consideration to those things that we feel are related to us. As the great conservationist Aldo Leopold argued, "all ethics so far evolved rest upon a single premise: that the individual is a member of a community of interdependent parts."[2]

Therefore, a key step in expanding our moral sphere to the natural world is to demonstrate our relationship to it. In his famous essay, "The Land Ethic," Leopold argues that we need to begin to view the land as a community to which we belong and upon which we depend. The insights of evolution and ecology tell us that we are defined by, depend on, and are inseparable from the land, if not all of nature. As much as any king, queen, or president, the land itself has played a decisive role in the history of our civilizations and our nations. Therefore, we ought to realize that it is logically consistent to include the land and other species within our moral community as well.

Yet, as discussed above, important and influential strains in Western religion and philosophy have held that humans are separate from nature. If all ethics assume a feeling of shared community, then an ethical obligation to biodiversity is impossible if we insist on a fundamental separation of humans and nature. This is one reason why the idea of another species having "rights" (let alone entire natural communities) is so preposterous to many people.

However, Leopold was not bowed by the challenge of extending our ethical community—those to whom we have a moral obligation—to include the land. He recognized that, throughout human history, we have extended our ethical community many times. For example, as he points out in "The Land Ethic," twentieth-century Americans view slavery very differently than the ancient Greeks did (or, for that matter, than some Americans did only a few generations ago). Unlike the Greeks, we no longer hold the view that slavery is acceptable and that slaves are property that can be killed on a whim; in our culture now, such practices are morally reprehensible.[3]

What does this mean for a biodiversity advocate today? Clearly, we are a long way

---

right of both farmer and fisherman to use the Basin's natural resources, but it also insists on the land (and waters) of the Klamath as an entity with ethical standing as well. From this perspective, it is impossible to view the Klamath controversy as one that is simply about farmers versus fish. People do not stand apart from nature; our dependence on it requires us to acknowledge our obligations to the nonhuman world. Farmers and sucker fish are both members of the community of life in the Basin, and the interests of any one member of the community do not simply supersede the interests of the community as a whole. Instead, the human stewards of the Klamath Basin have a responsibility to ensure the health of both the human and natural communities of the Basin and to protect their ability to maintain and renew themselves.

COMMUNICATIONS TIP

Talking about Rights and Responsibility

When an opponent brings up "rights" (property rights, individual rights), that is an opening to talk about responsibility—both the responsibility people feel to protect the earth for the future and the responsibility they have for their own actions. Responsibility is a primary American value. Across society, the value of responsibility is shared strongly by almost all demographic groups.

Keep in mind that a message that relies only on guilt—constantly reminding people that species are dying off—without providing the connections and benefits of saving species and habitats is ineffective with most audiences.

from the day when Leopold's land ethic prevails as the standard we use to judge our uses of the land and our actions toward other species. And while it may be hard to persuade a county planner, a congresswoman, or most of the public that other species, or even ecosystems, have rights, it may not be so unintelligible to assert that we have responsibilities to the land and its nonhuman inhabitants.

One reason may be because, as we have seen, the language of *rights* presupposes a theory in which the interests of isolated individuals are pitted against each other. But when we emphasize our *responsibilities*, we necessarily invoke the interrelationships among those individuals (the community) and all the benefits that such a community of interdependence provides. Communities are defined by shared interests among their members, even if those interests are limited to having a stake in the health and productivity of the same geographic area. Based on this connection, we can expand the boundaries of our community of ethical concern.

Opinion research suggests that the notion of responsibility based on shared interest is intuitive for many people. In the Biodiversity Project's 2002 national biodiversity poll, people were asked why humans should bear responsibility for habitat protection and loss. The messages that resonated with the greatest number of people were those focusing on the connection between habitats and the services (clean air and water) and goods (medicines) they provide people, or those emphasizing that habitats are communities that include many species (habitats are home to "so many species of plants and animals"). The message that had the least impact was one that simply stated that people are killing dozens of species each day—in other words, a message that implied human culpability but did not emphasize the interdependence between people and the natural world.

The language of responsibility clearly resonates with the public. At a time when our culture seems preoccupied with individual rights and self-interest, talking about responsibility reminds us of our obligations to each other and offers a way to embrace the natural world that is our community too.

Notes

[1] For some good examples of traditional atomistic conceptions of human rights, see the following United Nations Declarations: "The Universal Declaration of Human Rights" (http://www.un.org/Overview/rights.html), "The International Covenant on Economic, Social, and Cultural Rights" (http://www.unhchr.ch/html/menu3/b/a_cescr.htm) and "The International Covenant on Civil and Political Rights" (http://www.unhchr.ch/html/menu3/b/a_ccpr.htm)

[2] Aldo Leopold, *A Sand County Almanac: With Essays on Conservation from Round River* (New York, NY: Ballantine Books, 1966), 239.

[3] Ibid, page 237.

[4] Brad Knickerbocker, "Drought and a Western" *Christian Science Monitor* (May 24, 2001).

[5] Ibid.

Obligations to the Future

by Daniel Swartz

Introduction

While the social contract may place obligations on us to acknowledge the rights of others in the present, what are our obligations to others in the future? Do we owe anything to people who have not even been born yet? Ancient religious teachings framed moral questions in the very long term. Today, these teachings can help us understand ways to think about our obligations to the future and the importance of a long time-frame for environmental decision making.

Covenantal Responsibilities

Perhaps no statement better captures the essence of our obligations to the future than Deuteronomy 30:19: "I call heaven and earth to witness against you [plural] this day. I have set before you [singular] life and death, blessing and curse. Choose life, that you and your descendants might live."[1] Our choices today can affect the very survival of those yet to come—or at least surround them with the bountiful blessings or countless curses. Heaven and earth are called to witness this covenant precisely because it is to be eternal and because, when considered in an eternal time frame, the consequences of our actions extend across the Earth unto the heavens above. Rabbinic commentaries have understood the transition from plural to singular to have significance as well: while this passage is addressed to the entire community, it is an obligation to each and every individual.

This passage, far from being an isolated example, is only one of many that lie at the core of a great deal of religious thinking. In the Hebrew Bible, God makes promises that extend "l'dor va-dor," from generation to generation. Four hundred additional times, God speaks of covenantal responsibilities that extend to eternity (l'olam) or to eternity and beyond (l'olam va-ed).

KEY POINTS

Many ancient religious teachings call on us to consider the long-term consequences of our actions.

Our obligations to the future entail thinking preventively, not just for ourselves, and taking precautions against actions that raise the risk of likely, large-scale or irreversible dangers.

* Thinking long-term is seen as a fundamental part of being "good." Proverbs teaches (13:22), "a good person bequeaths to their children's children."

> Since God is also understood to be eternal, the future is just as imbued with presence, value, and meaning as the present—and so to discount the future is to deny the sovereignty of God. Living unsustainably then becomes a crime both against God and against generations yet to come.

Thinking long-term is seen as a fundamental part of being "good." Proverbs (13:22) teaches, "a good person bequeaths to their children's children."

Long-Term Thinking in World Religions

Long-term thinking is an integral part of many religious traditions around the world. The Iroquois believe in making decisions based on consequences for seven generations. The prime goal of society, according to Confucianism, is the moral elevation of the generation to come. A Hadith, a teaching traditionally ascribed to Mohammed, puts this in concrete terms. It says, "upon death, a person's good deeds will stop, except for three, a charitable fund, knowledge left for people to benefit from, and a righteous child."[2] Thus, thinking long-term means not only raising righteous children, grandchildren, and generations to come, but also enabling them to live in a world filled with knowledge and possibility.

For monotheistic religions, long-term thinking is rooted in both fundamental conceptions of God and commandments to pursue justice. God's presence is understood to be both unifying and universal. In terms of space, this dictates a shift from a "not in my backyard" mentality to a "not anywhere" mode. (Not incidentally, this also blurs the line between the public and private in many religious traditions. For example, in Judaism, humans are never considered to have true ownership, especially of land.[3] God is the only owner, and so there is no distinction between "private property" and "the commons," except that societies have greater obligations to institutionalize protection of common areas.) Since God is also understood to be eternal, the future is just as imbued with presence, value, and meaning as the present. Hence to discount the future is to deny the sovereignty of God. Living unsustainably then becomes a crime both against God and against generations yet to come. Since all people, in present and future generations, are considered equal under God, each generation should take care to use only its "fair share" of resources, in a sustainable fashion.

One of the most vivid teachings about the long-term consequences of our actions is found in early rabbinic commentaries (Genesis Rabbah) on the story of Cain's murder of Abel. God admonishes Cain (Genesis 4:10), "What have you done? The voices of your brother's bloods cry out to me." These commentaries explain that the unusual plural form of *blood* should be understood to mean all the potential descendants of Abel, all murdered through one act of violence on one person. In this context, acts that reduce biodiversity deny God's role as creator, kill countless generations of potential descendants of extirpated species, and rob the world and all who will inhabit it for generations of the presence of that species.

Taking the Future into Account: Precautionary Principle and Population

If societies took future generations into account, how might decisions and decision-making processes be different? First, truly valuing the future entails thinking preventively, with a good measure of precaution, for we can never fully know the long-term consequences of any action. In the Jewish legal tradition, prevention and precaution are rooted in Deuteronomy 22:8: "When you build a new house, you shall make a parapet for your roof, so that you do not bring bloodguilt on your house if anyone should fall from it." Because of lack of rainfall, roofs in the Middle East are typically flat, and they are used for gardens, laundry, or simply a place to feel the evening breeze. When one is on a roof, there is always a risk of accident—so homeowners have a responsibility to prevent accidents whenever possible. This not only marks a profound difference from a "buyer beware" philosophy, but also an acknowledgement that once accidents happen, it is often too late for healing or repairs. This "parapet principle," by the Middle Ages, becomes expanded to "anything that is potentially dangerous." (Shulchan Arukh, Hoshen Mishpat, Hilchot Shmirat HaNefesh, #427)

Second, danger itself is measured according to three criteria (see Rabbi Jacob Ettinger, Responsa Binyan Zion, 137): how "unreasonable" the risk is (that is, its scope), how likely some form of damage is, and how irreversible that damage might be. Thus, species extinction, for example, would be strenuously avoided, because it is permanent. Similarly, even if one could determine that the likelihood of risk to human or ecosystem health from genetically modified organisms is low, the scope and potential irreversible nature of harm would lead to an extra measure of precaution.

How does consideration of future generations and long-term consequences play out as religious traditions consider population growth? While religious teachings in this area are nuanced and complex, we should note that the command to humans to "be fruitful and multiply" (Gen.1:28) is by no means absolute; it comes, for example, after God has made identical proclamations to everything living in the waters or flying through the air (Gen. 1:22). Furthermore, many traditions look at the context surrounding population concerns. For example, early rabbinic writings recommended against procreation during famines or other times of limited resources, basing themselves upon an interpretation of the Noah story. (See, for example, Babylonian Talmud, Tractate Ta'anit 11a). In light of the way population growth, especially in high-consumption societies such as the U.S., affects the whole biosphere, we would not be fulfilling our obligations to the future if we did not further examine and make use of such traditions.

The contrast between religious timeframes that extend "from generation to generation" and present policies and actions in the U.S. and around the world is stark. When the future is considered at all—which is relatively infrequently—policy makers apply "discount rates" that assert that each succeeding year is worth less than the previous one. Within one generation, let alone across many, the future quickly becomes worthless. And in our day-to-day behaviors, ranging from automobile use to the waste of paper and resultant deforestation, we act, albeit often unwittingly, as if there literally will be no more tomorrow. Without deep changes in our behaviors, we are rapidly foreclosing options and passing on to generations to come a future that is devalued as we extinguish the tomorrows of countless species. May we hear the voices of their bloods crying out to us soon—and heed those voices to eternity and beyond.

Notes

[1] All biblical passages are taken from the Jewish Publication Society TANAKH Standard Edition.

[2] Azzam Tamimi, "Reflections on Islamic Political Thought, Past and Present," presented at Institut Catala de la Mediterrania, Barcelona, November 6, 2001.

[3] For an extended discussion of this, see Daniel Swartz, "Jewish Environmental Values: The Dynamic Tension Between Nature and Human Needs," in *To Till and To Tend: A Guide to Jewish Environmental Study and Action* (New York: Coalition on Jewish Life and the Environment, 1995).

Murky Waters: When There's No Clear Line between the Right and Wrong Choices

by Jane Elder

Introduction

One of the tragedies of being stuck in the midst of the sixth great extinction is that many of the decisions that individuals and societies must make to protect biodiversity don't come with a simple "right" or "wrong" label. Science can inform these decisions, but often it can't provide a clear prescription or simple answer, because the resolution requires difficult ethical choices where there is no easy "right" option. Our ethical and spiritual values can help guide us toward what we ought to do. Lasting biodiversity protection must remain responsive to people's values.

Pitting Ethical Choices Against Each Other

Biodiversity conservation at the beginning of the 21st century is fraught with ethical dilemmas and conundrums. This is happening in part because more flexible or desirable options have been foreclosed, or because the crisis is so dire that immediate action must be taken if there is any hope of a solution—such as those cases where not taking action has consequences that are more disastrous than the uncertain effects of taking action. Given the complexity of human values, the scale of some issues, and the unmovable fact that nature has no patience with human political decision-making timelines, we're in a fine mess.

The lack of a right answer means that sometimes the choice is the lesser of two evils, and people hate those kinds of choices. Often, conservation issues pit compelling ethical choices against each other. For example:

- Is it right to protect a dwindling whale population at the expense of the loss of timeless cultural hunting traditions?
- Is it right to capture the last individuals

KEY POINTS

Conflicts over biodiversity conservation often involve difficult choices between two or more "rights."

Complex questions about biodiversity protection cannot be answered by science alone—we need to apply values and ethical judgments to help guide our decision making.

of a wild species in order to breed them in captivity so as to prevent extinction?
- Is it right to displace or exclude indigenous or local peoples from biodiversity conservation areas?
- Is it right to tell a frightened village that they can't kill the tiger or the rogue elephant that is threatening their safety?
- Is it right to prevent commercial fishermen from harvesting threatened species when their livelihoods depend on fishing?

These ethical dilemmas are self-imposed by human culture (all six billion of us) at the beginning of a new millennium, and by the thousands of years of history that have brought us to this point. Often, those forced to grapple with the solutions had little to do with creating the problems. One could argue that we didn't see them coming. But here they are, and our generation is the one that must make the choices, difficult or not.

Taking Values into Account

These sorts of right v. right (or sometimes wrong v. wrong) choices often splinter traditional alliances when the debate is reduced to basic human rights v. long-term preservation of the biosphere. But of course it is almost always more complicated than this. Because issues of this scale are rarely resolved solely on the basis of scientific information, it is easier to grapple with difficult choices if we understand both the values and ethical systems that are at play and the ethical consequences of the various decisions. If we fail to address the ethical concerns of the affected parties the resolution will be temporary and shallow at best. Increasingly, biodiversity conservation will succeed only if advocates address the multiple cultural values that determine a successful solution for those directly (and indirectly) affected. People are much more likely to support a new policy, or to change their behavior, if they believe in their heart it is right, than if it was simply imposed upon them by an outside force.

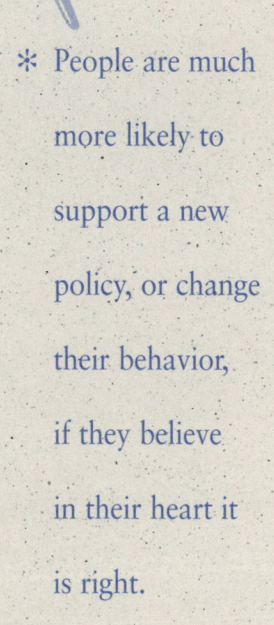

* People are much more likely to support a new policy, or change their behavior, if they believe in their heart it is right.

Right v. Right Conflicts: A Process for Ethical Decision Making

by Nancy J. Miaoulis

Introduction

Biodiversity is evidence of the ecological integrity of a place, and the continuing loss of species across the globe is a measure of the human impact on the biological world. If it is true that "ethics are central to our survival,"[1] we must at once take seriously the debate surrounding issues of biodiversity and humankind, assessing our responsibility toward the flora and fauna with which we were created.

Institute for Global Ethics—A New Paradigm

Often, the historical approach in many environmental debates is centered on the belief that there is one right answer to the ecological problems that plague our times. Both sides of an ethical equation cannot be correct, or so it is commonly thought. This approach, rather than providing solutions to environmental problems, has instead created impasses. As a result, many ecological debates end in stalemate.

The Institute for Global Ethics (IGE)—a nonsectarian, nonpartisan, global research and educational organization that promotes ethical behavior (www.globalethics.org)—asks us to consider the possibility that there are many "right" sides in the arena of environmental ethics, equally sound and deserving of consideration. IGE developed this "right v. right" paradigm (also called the Ethical Fitness™ process) "to help change the way we think about the world and to provide, through that change, a practical set of mental tools by means of which good people can make tough choices."[2] We offer the following case study and IGE's principles as an invitation to see how the Ethical Fitness™ process can help the parties in a right versus right conflict to exercise their moral capacity for reason, seek middle ground, and envision and act upon practical solutions.

Examining the Ethical Dilemma

The first step to resolving an ethical dilemma is to understand the conflict. Right v. right ethical problems are those that pit one right value against another. For example, in the Maine fisheries dilemma, the value of responsibility (for the economy and/or ones family) comes up against the value of respect (for biodiversity and the future of the marine species). To resolve these dilemmas, we need a framework to choose between the rights on both sides or to find the middle ground between them. The Institute for Global Ethics' decision-making model helps to identify the values in conflict and apply principles for resolution. It is a tool we can use to approach the complex ethical issues that surround environmental decision making.

Framing the Dilemma

Finding a way to discuss a dilemma with all interested parties can be difficult without having the tools to develop a common language. By applying an ethical lens through which to view an issue, a dialogue can begin. In the Institute's model, this process begins with framing the dilemma based on the values that are represented on all sides. The IGE has found that almost all ethical dilemmas tend to fit one or more ethical paradigms:

• *Truth v. Loyalty*—We learn as children that we should never tell a lie, but we also learn never to rat on a friend. Thus, we are taught the value of truth and the

---

KEY POINTS

There may be many "right"—or ethically legitimate—sides to conservation debate. Framing a conflict in a context of commonly understood ethical paradigms and principles can illuminate the values at the heart of the debate, clarify the perspectives and roles of the stakeholders involved, and point toward resolution.

value of loyalty. But what happens when a situation arises that asks us to be honest and loyal, and in choosing one, we compromise the other?

- *Justice v. Mercy*—Certainly the values of fairness and equity are essential in society, which is why we have rules and regulations that are meant to apply equally in every situation to every person. But what kind of world do we live in, if justice is applied so even-handedly that compassion and caring do not enter the equation? It is right to stick to the rules, but it is also right to be compassionate.

- *Individual v. Community*—This paradigm can also pit the needs of the small group against the needs of the large one. Is it always best to make the choice that respects you and your loved ones at the risk of the larger community? Or is it sometimes best to look at the bigger picture and try to assess the impact that your decision will make for others involved?

- *Long-term v. Short-term*—We know from operating our own household budgets that it is not always easy to determine whether we should pay for something to meet a present need or save for a future endeavor. The value of meeting needs that are current versus the value of conserving for needs in the future is one of the most common conflicts in environmental debates.

Analyzing a dilemma according to these paradigms can reveal the values at the heart of an ethical conflict and can, therefore, inform and help pave the way for an acceptable and equitable solution. In the example of the fishing industry debate, it is important to identify whose dilemma this is. In this case, there are a number of players and each one has somewhat different interests. The judge faces the dilemma of whether to set further restrictions on the ground-fishing industry or not. It is right

ETHICS IN THE REAL WORLD

A Clash of Values: The Maine Fisheries Dilemma

A recent federal court ruling has inflamed yet another classic conflict—this one between the economic needs of commercial fishermen in the Gulf of Maine and the future of rapidly diminishing groundfish stocks (such as cod and haddock, historically two of the most important fisheries in New England). Populations of cod, flounder, and other fish have crashed precipitously in recent decades. In addition, these fishing industries take a heavy toll on other species harvested in their nets (including marine mammals such as harbor porpoises). Adding additional restrictions to an earlier mediated agreement, a federal judge closed most of the Gulf of Maine to fishing for the months of May and June and declared thousands of square miles of ocean off-limits to fishing year-round. The ruling also cut back on the number of fishing days each boat is allocated, which reduced many fishermen's allowable days at sea by 50% to 70% and denied those with latent permits the right to fish at all. Furthermore, the ruling requires fishermen to invest in new fishing gear and to fish in more confined areas. The fishing industry believes that the standards set by the court will make it difficult for the 200 boats of Maine's groundfish fleet to stay in business.

※ In the end, it comes down to the values we commit to and the lens of ethics we apply to situations. All the regulation in the world won't preserve biodiversity. It will have to be about something much more meaningful and lasting.

on the one hand to protect the ground-fish and allow them to regain their numbers, thereby promoting the health of the ecosystem to which these fish are integral. It is right on the other hand to allow fishing to continue with minimal restrictions for a small industry that has invested heavily in previous compliance efforts as a sole means of supporting its workers' families and contributing to their communities.

Of the four dilemma paradigms explained above, the two that seem most appropriate to this situation are individual versus community and long-term versus short-term. If one sees the individual as the fisherman and the community as the marine ecosystem or the other humans on earth that rely on that system, the rights on

Communicating About Right v. Right Conflicts: An Interview with Abby Kidder, the Institute for Global Ethics

Abby Kidder, Senior Education Associate, has been with the Institute for Global Ethics since its inception in 1990. Among other interests, Abby has expanded the Institute's newly evolving environmental work. She has recently written a secondary school curriculum called "How Big is Your Backyard? An Ethics-Based Approach to Environmental Decision Making," published by the Institute for Global Ethics.

Do the principles you've suggested for addressing right v. right conflicts have implications for an organization's approach to communications? Absolutely. The Institute's process provides a common language for all parties to use in communicating about the issues at hand. Instead of leading us down the path of who is right and who is wrong, it gives us a framework for constructive dialogue around what are often controversial issues. If we can begin to see the questions of biodiversity as right-versus-right instead of right-versus-wrong, we have gained a platform for discussion that would not otherwise be there. If, as an organization, you can frame the ethical debate, outline the "rights" on both sides, and legitimately weigh various resolutions, you will draw support from all interested parties instead of alienating those whom you most need to reach.

What advice would you give to a group that is communicating to decision makers, the public, or even the other side in a right v. right controversy? Be sincere and do your homework. Find out what makes the particular issue an ethical dilemma, and do enough research to gather the pieces from all sides. Issues around biodiversity are complex and often involve many players. It's not about watering down the decision to black and white so you can prove the other side wrong. It's about understanding the ethical values that shape our culture and figuring out what to do when those values come into conflict with each other. There has to be a genuine interest in finding resolution that is honest and compassionate. This kind of sincerity and professionalism are part of putting values into action and move ethical decision making away from mere discussion to necessary practical action.

Does being in a conflict where the other side also has legitimate ethical claims affect the way you should communicate? Are there any added ethical responsibilities and considerations you should take into account when you are representing this kind of situation to the media? Again, the basis of the Ethical Fitness™ process is to identify the common values on all sides and move away from "right v. wrong" labeling to a respectful and constructive conversation. When communicating with the opposition or the media, you can begin by recognizing the ethical nature of the issue, stating the "rights" of each interest, and focusing on finding resolution based on shared moral values. In the end, it comes down to the values we commit to and the ethical lens we apply to situations. All the regulation in the world won't preserve biodiversity. It will have to be about something much more meaningful and lasting.

both sides are clear. Similarly, it is right to have fishermen support their families and their livelihood for the short-term but it is also right to preserve the future of the ground-fish stocks for the long-term.

Moving toward Resolution

After gathering the relevant information, clarifying the issues, and determining the paradigm(s) that the dilemma best fits, the next step is to apply a set of principles that point toward a resolution, a decision that will be based on choosing the highest right under the circumstances. The resolution principles put forth by the Institute for Global Ethics are based on philosophical traditions and provide a useful framework for determining an action to take:

- *Ends-based thinking (utilitarian):* This is often thought of as a decision that considers the "greatest good for the greatest number." What decision would the judge in this case make if she were doing what benefited the largest number? Probably she would decide to impose the restrictions, because even the fishermen and their families and their communities don't add up to the number of living organisms affected when a system is over-fished.
- *Rule-based thinking (the categorical imperative):* This decision considers the precedent that you would set by making a decision. If everyone who came behind you made the same decision you are about to make, would you be comfortable with the decision? What decision would the judge make if she were setting a precedent for generations to come? Again, she might set the restrictions to protect a limited and fragile resource.
- *Care-based thinking (Golden Rule):* If you put yourself in the other person's shoes, what would you do? What decision would the judge make if she were making a care-based decision—one where she had to consider what it would be like to be in the fishermen's shoes?

She would probably choose to alleviate the restrictions and promote the stability of the industry and the health of the fishing communities.

Beyond these three decision-making principles, there is also the option of finding a third way out—a choice that doesn't have to be "either/or" but can be a creative solution that finds middle ground. While not all dilemmas lend themselves to a third option, it is always worth searching for one.

No Easy Answers

The IGE process does not provide easy answers. Instead, it provides a common language to communicate about the complex issues that arise when our core values come into conflict. The framework offered here is a way for us to begin to make sense of environmental dilemmas as moral, self-reflecting agents. Our use of such a framework and the resolutions we arrive at will be determined by our capacity to understand all sides of an issue, communicate clearly, and make decisions based on the values we all share. As Rushworth M. Kidder writes:

> The more we work with these principles, the more they help us understand the world around us and come to terms with it…In that act of coming to terms with the tough choices, we find answers that not only clarify the issues and satisfy our need for meaning but strike us as satisfactory resolutions…and little wonder that as we practice resolving dilemmas we find ethics to be less a goal than a pathway, less a destination than a trip, less an inoculation than a process.[3]

Notes

[1] Rushworth M. Kidder, *How Good People Make Tough Choices: Resolving the Dilemma of Ethical Living* (New York: Fireside Books, 1995), 8.

[2] Ibid, page 9.

[3] Ibid, page 176.

> ✳ This process does not provide easy answers. Instead, it provides a common language to communicate about the complex issues that arise when our core values come into conflict.

How Shall We Live?
Environmental Equity and Justice

by Daniel Swartz

Introduction

How much we use—how we choose to live—has direct consequences on other species and the health of the environment. But lifestyle also has consequences for equity and justice, consequences that religious traditions have reflected on for thousands of years. We cannot effectively address the biodiversity crisis until we address disparities in wealth and power that drive human exploitation of the environment and other people. Acting justly toward nature and each other is not just good for other species or the poor and vulnerable. As the prophets and teachers of the Judeo-Christian tradition have warned, it is also the only way for the rich and powerful to save themselves.

Widows, Orphans, and Strangers: Prophetic Calls for Justice

People trying to promote sustainable living sometimes discuss the concept of the "ecological footprint." Use fewer resources, and your footprint is smaller; use more, and it is larger. And because the earth is itself limited, only so many large footprints can fit on it.

The prophet Ezekiel used terms not far distant from modern notions of ecological footprints. He wrote, "Is it not enough for you to graze on choice pasture, but you must also trample with your feet what is left from your grazing? And is it not enough for you to drink clear water, but you must also muddy with your feet what is left? And must My flock graze on what your feet have trampled and drink what your feet have muddied? Assuredly, thus said the Lord God to them, Here am I, I am going to decide between the stout animals and the lean" (Ezekiel 34:18-20).[1]

For millennia, prophetic voices in religious communities have been advocating on behalf of the "lean." In the Hebrew Bible and the Koran, the lean or the vulnerable are personified as the widow, the orphan, or the stranger. All have little power in many traditional societies, and they may even be viewed as signs of ill fortune or the enemy. The Gospels add such social outcasts as prostitutes to the list of the "least of these." Other traditions may speak less of justice, but they reach out to the vulnerable in other ways. For example, the Hindu tradition teaches "daya," compassion toward all that is focused on those with the least power. Traditionally, prophetic justice has focused on ending oppression and economic injustice. Increasingly, however, the voices of Ezekiel and his compatriots are being heard in the realm of environmental issues, in what has become known as eco-justice or environmental justice.

Environmental Justice: A New Golden Rule?

What is environmental justice? In the broadest sense, environmental justice examines how issues of power, equity, and opportunity play out in the realm of the environment. That is, do the powerful consume more resources? How does such consumption affect the poor, the voiceless, and the vulnerable? Who—and what—suffers most from environmental degradation and health hazards? Who benefits and who pays the cost from human activities that affect the health of the environment? And what does the practice of environmental injustice do to one's well-being, one's soul?

According to the World Resources Institute, the United States and 23 other

---

KEY POINTS

Judaism, Christianity, Islam and other major religions declare that we have a duty to care for the poor and vulnerable, a message that underscores the need for just and sustainable solutions to the economic inequities that, in many cases, drive biodiversity exploitation.

Religious traditions warn about the dangers that over-consumption and environmental injustice pose not only to the poor, who are denied equity, but also to the powerful who perpetuate injustice and fall victim to the "idolatry" of materialism.

industrial nations, with just 16% of the world's population, generate 40% of its greenhouse gases and 68% of its industrial waste.[2] And within the industrialized world, people in the U.S. rank at or near the top in almost every category of natural resource consumption—using, for example, twice as much fossil fuel as the average resident of Great Britain, and producing three times as much waste as the typical Western European. Despite having just 5% of the world's population, the U.S. consumes one-third of the world's paper, leading to loss of vital forest habitats not only here but also across the globe.[3]

What guidance can religious traditions give us in the pursuit of justice and equity, and how might a vision of environmental justice change how we live, not only as "grazers," or consumers, but also as human beings living in relationship to a complex, diverse biosphere? While the links between the consumption of the powerful, the poverty of the vulnerable, and the destruction of the environment may be more or less obvious, this much is clear: Ezekiel's admonition would be directed against us today. The coffee we drink, in most cases, is a product of deforested habitats and industrial plantations; the hamburger we eat comes via trampled lands. At the other end of the spectrum of wealth and power, poverty and environmental degradation are often linked in an escalating spiral. The poor are driven to marginally arable lands, where, out of desperate attempts to survive, they degrade the land further, leading to more poverty, and so on.

In this country, a number of studies, from such diverse sources as U.S. Environmental Protection Agency and the United Church of Christ, have found that disproportionate numbers of landfills, incinerators, chemical plants, and hazardous waste sites are situated in low-income communities and/or communities of color.[4] Lead poisoning affects African Americans and the poor at rates almost an order of magnitude higher than those of middle-class whites.[5] Obviously, chemicals themselves do not discriminate, but communities with less power to protect themselves have disproportionate environmental risks thrust upon them.

Environmental injustice does not occur only between different parts of society or different nations. If the "field mark" of injustice is might becoming right—the powerful getting benefits at the expense of the vulnerable simply because the vulnerable are less likely to fight back—then all of our relationships with the world around us need to be examined through the lens of environmental justice. Who, or what, are the "widow, orphan, and stranger" today? Are they trees that have no legal standing? Are they species whose existence stands as a roadblock to profits?

✲ We cannot effectively address the biodiversity crisis until we address disparities in wealth and power that drive human exploitation of the environment and other people.

Results from the Biodiversity Project's 2002 national biodiversity survey suggest that more people in the U.S. are willing to take action in their own lives to protect the environment, but many still don't make the connection.

• Approximately six in ten (59%) Americans say that in the past year they have changed what they do as a consumer by paying a little more for products that are friendlier to the environment.

COMMUNICATIONS TIP

Americans often do not make the connection between their lifestyle choices and the impact those choices have on biodiversity, but advocates can help them see this connection by pointing out the implications of our choices and by promoting personal actions directly linked to protecting the environment and human health.

Here's an example of a message that helps make the connection between personal actions (reducing consumption), the environment, and social justice:

"The traditional American Dream once focused on greater security, opportunity, and happiness. Increasingly, that dream has been supplanted by an extraordinary emphasis on acquisition. The recent commercial definition of the American Dream has hidden costs for the environment, our quality of life, and our efforts to create a just and equitable society. . . . If we wish to reverse this trend and preserve necessary resources for our children and future generations, we must shift and reduce our consumption of resources." —Center for a New American Dream (www.cnad.org)

In a variation on the "Golden Rule," Rabbi Hillel wrote, "What is hateful to yourself, do not do to your fellow creatures." The ambiguity in Hillel's use of the word creature suggests how we can turn this rule into an ethics- or faith-based environmental impact assessment of a particular environmental question: are you doing something "hateful" to your fellow creatures; and, are you including non-humans when you consider "fellow creatures"?

Coins Before Our Eyes

Finally, religious traditions can also help us realize the cost to the powerful of environmental injustice. Most faiths teach about "moderation" in one form or another, about the importance of spiritual discipline and self-limitation. For example, Christianity speaks of the virtues of humility and generosity and the vices of greed and covetousness. As we consume more and more, we aggrandize ourselves at the cost of our relationships with others, at the cost of the lonely fate of the narcissist. We even, in one sense, lapse into idolatry, worshiping the "stuff" we have made with our own hands or bought with our labors. The words of Isaiah ring as true today as ever, especially if we replace "chariot" with "SUV." Isaiah wrote, "Their land is full of silver and gold, there is no limit to their treasure. Their land is full of horses; there is no limit to their chariots. Their land is full of idols; they bow down to the work of their hands, to what their own fingers have wrought" (2:7-9). This materialistic idolatry, this worshipping of wealth and power, not only distances us from God and from the less powerful, it also dims our eyes to wonder. As Nachman of Bratzlav taught, "the smallest coin held before the eyes can hide the grandest mountain."[6] To pursue environmental justice and to preserve biodiversity, we need to cast coins out from before our eyes and open them wide.

Notes

[1] All biblical passages are taken from the Jewish Publication Society TANAKH Standard Edition.

[2] World Resources Institute, "Earthtrends 2001," http://www.earthtrends.wri.org.

[3] World Resources Institute, "Earthtrends 2001," http://www.earthtrends.wri.org.

[4] United Church of Christ, Commission on Racial Justice, "Toxic Waste and Race in the United States: A National Report on the Racial and Socio-economic Characteristics of Communities with Hazardous Waste Sites" (New York: United Church of Christ, 1987).

[5] Ibid.

[6] Rabbi Nachman of Bratzlav, "Likkute Mohoran" (Lessons of Rabbi Nachman), #35, section 5.

SECTION IV

Thinking Locally, Acting Globally: Steps toward an Ethic for the Biosphere

THINKING LOCALLY, ACTING GLOBALLY: STEPS TOWARD AN ETHIC FOR THE BIOSPHERE

Saving Land and People

✴ And why talk about American values in a discussion of biological diversity? Because we recognize that saving land alone is not an enduring solution to the diversity crisis.

by Peter Forbes

Introduction

Along with the Wilderness Society, the Trust for Public Land (TPL) has been leading the way in incorporating ethics and values into the day-to-day work of protecting the land. The result is a shift in TPL's mission and vision—from "dollars raised and acres saved" to finding ways to deepen people's connections to the land and each other. TPL's explorations have helped spark a new discipline called Conservation Sociology, that examines what the land means to people, the different ways people value nature, how to have a healthy relationship to the land, and how to foster caring about the environment. By reaching out to people of diverse cultures, TPL is encouraging the connections among conservation, ethical behaviors, social change, and healthy communities.

Connections to the Land

Speak these American place names to yourself: Androscoggin, Bear-Paw, Bitter Root, Black Canyon, Catawba, Blue Ridge, Congaree, Elkhorn Slough, Hollow Oak, Jamaica Bay, Keweenaw, Gathering Waters, Lummi Island, Mojave, Otsego, Prickly Pear, Siskiyou, Tecumseh, Wolf River. These places, and others like them, speak of our history. They are the waters, the food, the wood, the dreams, and the memories that form us. When we care for the places that define us, we give ourselves a gift of memory and connection. These are the places that inspire our belonging, replenish our souls, and remind us that where we live is like no other place in the world.

As we fail to protect and hold our special places, every place in our country begins to look like every other place. Sprawl is gobbling up the U.S. countryside at 365 acres per hour.[1] More than 3 million acres of forests, farms, and wetlands will be paved over this year alone. There are more malls in America today than high schools.[2] These changes in the landscape change who we are, and the loss of our bearings continues every day across America: the loss of the little places we knew as children, the landscapes that told us we were home, the felling of a forest, the disappearance of the grizzly bear and the wolf, the vanishing ways of life. *"The world I knew is gone."* In fact, many of the places we once knew *are* gone, or are rapidly disappearing. Our health and security as a species are based on the integrity of our relationships with all the life around us. What we are losing are the deep relationships we have had with land and through land, with community and with a story much bigger than ourselves.

In losing our connection to the land, we lose an important source of information about how we might live differently. Increasingly, there is only one story to hear and one story to tell. For too many, this has become a world where the point of trees is board feet, the point of farms is money, and the point of people is to consume. We are teaching our children that the only story that matters is the one playing in their heads, the only land that matters is what they might own, the only people who matter are themselves, and the only time

DOROTHEA LANGE PHOTO COURTESY OF USDA NRCS

78 **Biodiversity Project** *Ethics for a Small Planet: A Communications Handbook*

that matters is now. We have created self-centered humans, and they are endangering the long-term health of their own species and the long-term health of all life on this planet.

The Promise of Land Conservation

Aldo Leopold said, "there are two things that interest me: the relationship of people to each other, and the relationship of people to the land." Fifty years after Leopold's death, conservationists are still struggling to re-think the promise of land conservation as a force for social reform that creates a relationship between people and the land that positively transforms both. Many now practicing land conservation aspire to expand the idea of conservation biology to include a commitment to protect the natural habitat required to nurture responsible, joyful human beings—who in turn will delight in protecting the environment around them.

To expand the success of land conservation in the U.S. we must learn to speak of it not in scientific terms but in terms of human values and human lives. Why talk about U.S. values in a discussion of biological diversity? Because we recognize that saving land alone is not an enduring solution to the diversity crisis. Many of us are skeptical about the lasting value of holding actions like saving land, unless we also take on the messiness and complexity of struggling for the soul of our country. We have faith that the act of conserving land, and conserving our human relationships to the land, has the power to transform the choices people make in their daily lives.

Behind every protected national park, biosphere reserve, wildlife refuge, and endangered species is a powerful story of human restraint and forbearance. At its best, land conservation offers an alternative vision for the future. It harnesses the mystery, the unspoken love, that brings people out of their homes to protect the farms, rivers, and mountains of their lives.

Such favorite local places don't necessarily contain any known threatened species of plant or animal, but their loss would mean *an extinction of human experience*. Land conservation can tear down the walls that divide people from themselves, from one another, and from nature, and thus can become the starting point for a renewed community life. Conserving land can bring into people's moral universe a renewed sense of justice, meaning, respect, joy, and love, and make people feel more complete.

Stories of Hope

Quleana: Land and Responsibility
Glenn and Kathy Davis are learning Hawaiian traditions and rebuilding Hawaiian culture by bringing young Hawaiians back to the taro fields. For thousands of years until the 1960s, taro was a primary food on the islands, and it was grown everywhere in irrigated, terraced fields. The culture developed with this plant. The Hawaiian word for "family," *ohana*, is also the word for the taro root. *Quleana* means both "land" and "responsibility." But since the 1960s, many Hawaiians have left taro farming and the traditions of their ancestors for jobs in tourism. Recent conservation efforts have enabled some of them to return to the abandoned taro fields and restore the values that strengthen them as a people. Glenn told me, "people are beginning to remember that we are organic and part of the land.... Now that we've really come back and are committed to the taro again, there are more birds singing in the jungle. The water is flowing again. We have come home."

To Ride across the Land
The return of the Nez Perce to their ancestral grounds in Wallawa County, Washington, is a story of cultural and personal healing. Through a partnership with the Trust for Public Land that bought for them a 10,000-acre ranch along Joseph Creek, the Nez Perce returned to a region from which

they had been forcibly removed over 120 years before. For individual Nez Perce, the return to Joseph Creek offered a step toward a more integrated way of life. Allan Pinkham, a member of the Nez Perce tribal council, told me:

> "Returning to this land allows us to practice being good neighbors again. Our neighbors are the salmon and the eagle and the wolves, and, yes, particularly the white ranchers and even their ancestors who killed our ancestors and drove us off this land. The land teaches how we must all live together as good neighbors. So to be more self-sufficient, to be more our selves, we have chosen to rebuild our lives around the land. This will make us stronger and give us back our self-respect that no one can then take away. Yes, we need schools and jobs. But the best way for the Nez Perce to fight drug-abuse and alcoholism is to restore the salmon, and to bring back the wolves, and to ride across the land on our own Appaloosa."

Best Intentions
In a neighborhood in Colorado Springs, a small handful of dedicated people came together to say they cared about one piece of land—Stratton Ranch. By halting the building of luxury homes on land they used together and loved, these neighbors saw the future, not in terms of economic growth, but in terms of community. Their connection to this last piece of open land in a developing area moved them to protect it, and their connection to the land connected them to one another. Richard Skorman, who was later elected to the City Council, said, "I had given up on this city, thought we just had caved into the whole argument that anything valuable and worth doing was just plain impossible. Saving the Stratton land was a big step in saving ourselves. We can do good. It's there everyday now, reminding us of our best intentions as a community."

Changing Who We Are and How We Live
There is a transforming power in land conservation—protecting what we love can change who we are and how we live.

Classie Parker is at the center of such a transformation. She lives in Harlem, just a few blocks from the hospital where she was born. For many years, Classie felt stuck on a street where no one knew anyone else and drug dealers ran everything. In 1992, Classie's apartment building stood next to a 3,600-square-foot vacant lot that was crowded with crack vials, needles, abandoned cars, and garbage of every kind. When Classie got the idea to create a garden on that lot for her father to work in, she recruited her brother as well as a Hispanic couple who lived nearby and their five children to help her. Classie had a vision for a place where the old and young could work together. Today, the thriving garden is called "Five Star," in honor of the five adults and five children who started it.

An eight-foot-high chain link fence can barely keep the sunflowers from pouring out into 121st Street. With two large townhouses protecting either flank, the garden itself is bold and vibrant. A dozen vacant lawn chairs are organized loosely around leaning tables and empty crates as if a card game or a good meal had just been finished. There are rows of corn, plots of vegetables, climbing snap peas, grapevines, fruit trees, and a dogwood. One hears birds. Men and women of all ages hang on the chain link fence talking to friends on the street and then turn back into the garden with a hoe or a laugh.

Five Star is breathtakingly beautiful and heavy with life. It is stewardship and wildness wrapped together and dropped down on 121st Street. Classie produces food, beauty, tolerance, neighborliness, and a relationship to land for people throughout her part of Harlem, all on less than one-quarter of an acre.

What can conservationists learn from Classie Parker and Five Star Garden? From Glenn and Kathy Davis and the renewed taro fields of Hawaii? From Allan Pinkham and the return of the Nez Perce to their ancestral lands? And from Richard Skorman and the rescue of Stratton Ranch? What connects these acts of conservation? The answer is, all of these sites were protected because people had a relationship, a connection, with the land.

The enduring value of people's relationship to the land might best be measured by the extent to which that relationship evolves beyond self-interest. All healthy relationships entail sacrifice and are never solely about what makes one person feel good; instead they are also about what's good for someone else. Relationship implies a responsibility that goes beyond one's own dreams. Wendell Berry put it this way: "to grow up is to go beyond our inborn selfishness and arrogance; to be grown up is to know that the self is not a place to live."

Conserving Relationships
Viewing land conservation as the conservation of relationships forces us to become more self-aware because it poses difficult questions about our right and wrong behavior. How is our land use improving or degrading relationships between us and the land, and within the whole of the land community? What sort of natural habitats—from wilderness, to riparian corridors, to working landscapes, to city parks—do we humans need in order to have vital relationships, to be healthy, to act from our values, and to respect the needs of other species in the whole land community?

We need a seamless view of all life—people and nature together—because all healthy life needs a connection to other healthy life. People need a direct relationship to the forms of life that are found in working landscapes, in urban gardens, in millions of backyards, and in the truly wild. Furthermore, we must show forbearance to the rest of life: we must have the respect to leave many places completely alone.

The only way for a culture to stop viewing land as a commodity is to stop thinking of land objectively, as the other, and to begin thinking of land subjectively, as us. Through this new conceptual framework, our most valued natural areas might be seen as an expression of ourselves at our best. Wilderness areas could become our national symbols of respect, ancestry, and pilgrimage. Our strengths as a people might emerge from the quality of our relationships with the land, including our sense of care, well-being, neighborliness, trustworthiness, and health. There is no single branch of science or philosophical tradition to help us protect the relationships between people and the land. Unlike conservation biology, this work of protecting ways of life, or habitats for people, has no highly defined project selection criteria. But it includes growing healthy food, having safe parks and clean rivers accessible to people, building relationships with the land that inspire our sense of ethics and art, maintaining a culture of mutual aid and an appreciation of local beauty, defining our limits as responsible creatures, and protecting our cultural and ethnic diversity—all of which contribute to the health and well-being of all species on the planet.

> The tactics of the property rights side presents us with an opportunity to inject something entirely new and compelling into the mix....They give us an opening to talk about values, and by doing so, recognize each other as people.

Our success in conserving land will be measured not only by how much nature we can protect, but also by the amount of love and respect for the land we can engender in people. Our greatest achievement will not be to say, "we saved this place," but to say, instead, "you saved this place, you belong here, you are home."

A Great Remembering

I have heard others say that we are deep into a Great Forgetting; we are living in a time when our relationships are so fractured that we have almost forgotten why they were important in the first place. I prefer to believe that we are instead on the brink of a Great Remembering, a time when we can reconsider what matters most to us. Our relationship to land can be a reawakening to a more profoundly human life, a rekindling of what is most meaningful inside each of us, a way for humans to re-engage with the natural world.

Notes

[1] Natural Resources Defense Council, "Cities and Green Living: Smart Growth/Sprawl," http://nrdc.org/cities/smartGrowth/default.asp.

[2] Public Broadcasting System, "The Population Bomb – What Price," http://www.pbs.org/kqed/population_bomb/danger/price.html.

ETHICS IN THE REAL WORLD

Cease-fire at Pinkham Notch: Defusing the Wise-Use Movement

During the controversy over the fate of the Northern Forest in the 1990s, Bob Perschel, Director of the Land Ethic Program at The Wilderness Society, witnessed the power of values-based personal expression and honest dialogue. He saw first-hand how talking about personal connections to place, sharing values, and understanding where someone else is coming from could transform a confrontation between wilderness advocates and an angry crowd of property rights activists.

In the following interview, Bob recounts his foray into what Peter Forbes describes as "conserving relationships" and what resulted from it.

Biodiversity Project: Bob, tell us a little bit about the situation you were in.

Bob Perschel: We were holding a meeting at the Appalachian Mountain Club facility at Pinkham Lodge, in the middle of the White Mountains in New Hampshire. About twenty or so property rights people showed up with painted vans and placards. They marched on the highway and then came inside. The group attacked us for allegedly trying to ruin their communities and trying to "greenline" the entire Northern Forest. They became louder and angrier. The conservation community was in danger of taking a major hit. But there was an opportunity as well.

Biodiversity Project: What was that opportunity?

Bob Perschel: The tactics of the property rights side presented us with an opportunity to inject something entirely new and compelling into the mix, enlarge the possibility of collaboration, and move to a new way of dealing with the issues. They gave us an opening to talk about values, and by doing so, to recognize each other as people.

Many of the people who think of themselves as property rights advocates do so only because this issue was the first thing that came along that appeared to speak to their core values. But they have other values that can be awakened, if they can be moved away from their affiliations for a moment. When that small amount of room is created, it is time to introduce values that represent the higher aspirations of the human spirit.

Biodiversity Project: What were you thinking before it was your turn to address the crowd?

Bob Perschel: People were feeling unrecognized and this was fueling their anger. I decided that three things needed to happen. First, they needed to be recognized as individuals with a point of view. Second, the debate had to shift from a confrontation between the vague ideologies of groups to a dialogue between individuals. And third, it was critical that a new set of values be presented to them in a way that was neither offensive nor didactic. If this was done, each individual would be free to look within themselves to find their own representation of these values. From there, we could talk about the Northern Forest from a new perspective.

Biodiversity Project: How did you do this?

Bob Perschel: I spoke very slowly in an attempt to demonstrate that I was choosing my words with great thought and care. I said, "I know that not all of you will agree with what I have to say, but I want to thank you for listening to me." Something changed in the room. We now all had an individual sense of identity, even though we recognized that we didn't all agree. From that place it was truly possible to listen to each other for the first time.

Then I said, "For me to answer your questions, I think it is important that you understand who I am and what I do." I told them that I had been a forester and worked in the woods for many years, and that I believed that we needed both wilderness and a healthy forest industry.

Then I described how, during my time as a forester, my perception of my responsibilities changed. I became aware that we might lose 20% of all the species of life on the planet in the next 30 years. This possibility burned in my mind. I thought of my legacy to my son and all our legacies to future generations. What would my son say to me in 30 years? "Dad, what were you doing? What could you have been thinking?" I told the crowd: "I cannot accept the possibility of this loss. I cannot allow this to happen, not on my watch." I also told them: "I can't accept that we have to choose between protecting the environment and protecting jobs. There is a way to do both."

Biodiversity Project: How did the property rights demonstrators react?

Bob Perschel: What was interesting was that in the end I never had to answer the question originally put to me about "greenlining." No one ever asked me about it when I was done. There was no longer any need to debate this or seek conflict. Everyone in the room was, at least for a short time, beyond the conflicts that had so engrossed us.

Biodiversity Project: Did you see any long-term outcomes from this encounter?

Bob Perschel: Two things come to mind. First, this experience was really helpful to me when I accepted the assignment at TWS to figure out how we can more actively integrate and promote a land ethic in our work. It occurred to me that this work is subtle and actually has several components to it: evoke, define, legitimize, and apply. At Pinkham I was able to reawaken our common connection to place, which elicited a new response. The nature of our dialogue was defined when I spoke of my own deeply held beliefs and mentioned my family. I used my background as a forester to legitimize what I was saying. What was most instructive was that I did not move to the "apply" stage. This demonstrated that at any point in time activists could work on each of these components separately, in combinations, or as a whole. They provide an overall framework for the work of fostering a new American land ethic.

The second long-term outcome is related to the "apply" stage of the framework. Just this year New Hampshire completed a deal to protect the largest remaining private parcel in the state—172,000 acres called the Connecticut River Headwaters. The real challenge that day at Pinkham wasn't to agree or disagree on "greenlining," but to work together to come up with a sense of what shared and deeply held values we could draw on to build a land ethic that would help us make decisions about forest protection. The recent deal to protect the Headwaters didn't include everything that I would have liked to see. But we have come a long way, we are moving in the right direction, and we are still talking to each other and exploring what these beautiful places mean to us and what our responsibilities to them might be. I think that is progress. (From a telephone conversation with the Biodiversity Project and Bob Perschel, 15 July, 2002.)

The Earth Charter: Guide to a Sustainable Way of Life

by Dieter T. Hessel

THINKING LOCALLY, ACTING GLOBALLY: STEPS TOWARD AN ETHIC FOR THE BIOSPHERE

✷ The Earth Charter is a spiritually resonant, ethically coherent and socially practical expression of the transformative vision and eco-justice values needed to guide ecumenical Earth.

Introduction

By rapidly destroying the planet's inventory of life and altering nature to the point of "decreation,"[1] humans are creating dire ecological consequences that have global impact and impose severe social injustice. This historically new situation calls upon persons and social institutions on every continent to become world citizens who observe common ethical standards of sustainability, justice, and peace. In a world diminished by degradation of the natural environment and the impoverishment of people with little power, particularly in (what we might more accurately call) the two-thirds world, the common goal is to secure human well-being on a thriving earth.

Today, on six continents, environmentally responsive people—both religious and secular—are joining the "eco-justice" movement that seeks the well-being of Earth and people through human action for ecological well-being, environmental justice, and sustainable community.

The Earth Charter: a Global Ethic for Sustainable Living

The Earth Charter is a spiritually resonant, ethically coherent, and socially practical expression of the transformative vision and eco-justice values needed to guide ecumenical Earth. Completed in March 2000, the Earth Charter is a holistic, layered document that articulates the inspirational vision, basic values, and essential principles needed in a global ethic for Earth community. Its clear and universally valid imperatives give special emphasis to environmental challenges in a world that is getting hotter, stormier, less biodiverse, more crowded, unequal, and violent. But the Charter is more than "environmental"; it offers an inclusive ethical vision encompassing ecological integrity, human rights, equitable human development, and peace. The net result is a striking vision of global interdependence and a fresh and workable conception of sustainable earth community.

This document—now translated into many languages—has its origins in an international workshop on the Earth Charter sponsored by the Earth Council (formed after the 1992 Rio Earth Summit). Held in May 1995, at The Hague, it involved representatives from 30 countries and over 70 organizations. In 1997, the Earth Council joined with Green Cross International, led by Mikhail Gorbachev, to form the Earth Charter Commission. The Commission established an international drafting committee, which prepared, and received considerable feedback on, two benchmark drafts of the Earth Charter. Following a multi-year process of wide consultation with civil society groups on six continents, and after the drafting committee sifted numerous recommendations on ways to improve the text, the Earth Charter Commission approved and released the final version of the Earth Charter in March 2000, as a people's guide to Earth community ethics.

The body of the Earth Charter is comprised of four sections with the following titles:
- I. Respect and Care for the Community of Life;
- II. Ecological Integrity;
- III. Social and Economic Justice;
- IV. Democracy, Nonviolence, and Peace.

Section I states four broad, interdependent principles that we must commit ourselves to, and sections II, III, and IV discuss how to fulfill these commitments through 12 organizing principles. These 16 concise ethical principles (or norms) are of enduring significance and are widely shared by people of all races, cultures, and religions. They were composed with the intent to offer a common standard for the conduct of persons, organizations, businesses, governments, and transnational institutions to support a sustainable way of life. Moreover, the principles deliberately incorporate relevant points of consensus achieved in international agreements and at United Nations conferences in the 1990s.

The Charter's principles offer much more than meet the casual eye. The sub-principles in each section especially reflect input from civil society groups struggling for eco-justice around the world. Earth Charter principles also include key phrases drawn from already formulated international legal principles going back to the Universal Declaration of Human Rights. In addition, the principles include phrases from consensus language adopted in platforms of recent United Nations conferences aimed at resolving issues related to the environment, human rights, population, women, and urban development.

The main purpose of the Earth Charter is to establish a sound ethical foundation for our emerging global society, and to help build a sustainable world based on respect for nature, universal human rights, economic justice, and a democratic culture of peace. A less obvious accompanying intention of the Earth Charter is, in the words of the drafting committee chair, "to give to the emerging global consciousness the spiritual depth—the soul—needed to build a just and peaceful world community and to protect the integrity of Earth's ecological systems."[2]

Religious Resonance

The Earth Charter does not refer to God or a Creator, because some religions don't use these terms, and because the Charter's global worldview and ethic need to speak to agnostic and secular readers as well as religious ones. However, the Earth Charter contains spiritual affirmations about the evolutionary community of life and the human role therein that are congruent with biblical-monotheistic (and other religious) themes. The Preamble conveys the spiritual depth of universal interdependence and human responsibility within the Earth community:

> In the midst of a magnificent diversity of cultures and life forms, we are one human family and one Earth community with a common destiny. We must join together to bring forth a sustainable global society founded on respect for nature, universal human rights, economic justice, and a culture of peace. Towards this end, it is imperative that we, the peoples of the Earth, declare our responsibility to one another, to the greater community of life, and to future generations.

* Jews, Christians and Muslims can read the Preamble as an affirmation that Earth within the whole cosmos bodies forth the power, wisdom, and love of God. All of Earth community is valuable to God, who relates directly to and cares for the well-being of every kind.

Subsequent statements in the Preamble announce: "Humanity is part of a vast evolving universe. Earth, our home, is alive with a unique community of life." This statement reflects the influence of cosmologist Thomas Berry, who encourages us to see reality as "a communion of subjects, not a collection of objects."[3]

Jews, Christians, and Muslims can read the Preamble as an affirmation that the whole cosmos embodies the power, wisdom, and love of God. All of Earth community is valuable to God, who relates directly to and cares for the well-being of every kind. As the last sentence of the Preamble's second paragraph puts it, "The protection of Earth's vitality, diversity, and beauty is a sacred trust."

Later, the Preamble enjoins us to "live with a sense of universal responsibility" and to express "the spirit of human solidarity and kinship with all life" animated by "reverence for the mystery of being, gratitude for the gift of life, and humility regarding the human place in nature." The Preamble ends with a commitment to implement the Charter's vision and values in a sustainable way of life. Distinct threads of religious conviction, consistent with biblically informed faith and faith informed by sacred texts of other world religions, are woven into the tapestry of the Earth Charter's Preamble and Conclusion.

Earth Charter Ethics: a Brief Tour

The first of the Charter's four general principles in Section I—"Respect and care for the community of life"—affirms the interdependence and intrinsic worth of every kind of life. Such a holistic position moves beyond the Agenda 21 consensus reached at the 1992 Earth Summit that was anthropocentric and thus preoccupied with the "use value" of natural resources. From the foundational first principle flow three more general principles that specify shared human obligations: human responsibility for otherkind, i.e., "Care for the community of life with understanding, compassion, and love;" responsibility within and among human societies, i.e., "Build democratic societies that are just, participatory, sustainable, and peaceful;" and responsibility for future as well as present generations, i.e., "Secure Earth's bounty and beauty for present and future generations." Humans are to care for and to conserve the community of life sharing benefits and burdens. Enhancing quality of life and relationships —among people and with nature—are the essential aims. In other words, the human goal is to have and share life abundantly.[4]

The four general principles on Respect and Care for the Community of Life in Section I express broad commitments that then become operational in the twelve organizing principles the Earth Charter presents in Section II, Ecological Integrity; Section III, Social and Economic Justice; and Section IV, Democracy, Nonviolence, and Peace. Overall, the 16 ethical principles present a moral ecology of crucial values to embrace and necessary methods to follow in seeking truly sustainable development and community. These principles are common imperatives for the 21st century that are applicable everywhere and at all levels of moral agency—personal, institutional, and governmental. Below are some ethical features of Earth Charter principles 5 through 16.

Ecological Integrity

Section II, Ecological Integrity, leads off with principle 5, which summarizes the global/local agenda for preserving biological diversity by protecting and restoring ecological systems: "Protect and restore the integrity of Earth's ecological systems, with special concern for biological diversity and the natural processes that sustain life." Several subprinciples underscore the need for holistic development planning, management of renewable resources, and careful extraction and use of nonrenewables, within a larger strategy of conservation and habitat preservation. Observance of these subprinciples makes economic activity a subset of ecological responsibility, not the other way around.

Principle 6 highlights prevention—the "golden rule" of environmental ethics: "Prevent harm as the best method of environmental protection and, when knowledge is limited, apply a precautionary approach." The latter half of principle 6 invokes the precautionary principle as a practical guideline whenever knowledge is incomplete or inconclusive. The subprinciples specify what precaution really means, countering the tendency of business to obfuscate its meaning, in the same way that the term "sustainability" has been diluted and misused. All societies need meaningful precautionary measures to prevent more toxic pollution, global warming, ozone depletion, deforestation, aquifer and river degradation, and over-fishing.

Principle 7 emphasizes that population stabilization, reduced consumption, renewable energy systems, and full-cost accounting are related facets of an integrated approach to maintaining Earth's regenerative capacities while also seeking the well-being of human communities. The last two subprinciples under principle 7 enjoin us to "ensure universal access to health care that fosters reproductive health and responsible reproduction," and "adopt lifestyles that emphasize the quality of life and material sufficiency in a finite world."

Principle 8 calls for more scientific and technical research on sustainability, greater information sharing, respect for local and traditional knowledge and keeping vital environmental information (including genetic data) in the public realm.

Social and Economic Justice

Section III of the Earth Charter, Social and Economic Justice, continues to highlight eco-justice concerns by linking ecological sustainability with economic justice. Principles 9 and 10 present guidelines for overcoming poverty and achieving human development in an equitable and sustainable manner. Principle 9 emphasizes that eradicating poverty is an environmental as well as a social imperative. Its subprinciples call upon us to guarantee human environmental rights "to potable water, clean air, food security, uncontaminated soil, shelter, and safe sanitation"; and urge us to "empower every human being with the education and resources to secure a sustainable livelihood, and provide social security and safety nets for those who are unable to support themselves." Transparent, equitable, and sustainable international aid and trade is the ethical concern of principle 10.

Principle 11 affirms "gender equality and equity as prerequisites to sustainable development" to be achieved through "universal access to education, health care, and economic opportunity." Emphasis on this principle and its subprinciples regarding the rights of women and girls seems essential to any meaningful strategy of sustainable development. Nonetheless, the Earth Charter's stated commitment to full participation by women as equal subjects seeking full development in mutuality with men has prompted reactionary cultural and religious forces to ignore or attack this principle of a common global ethic.

Principle 12 reads, "Uphold the right of all, without discrimination, to a natural and social environment supportive of human dignity, bodily health, and spiritual well-being, with special attention to the rights of indigenous peoples and minorities." One of its subprinciples also affirms "the right of indigenous peoples to their spirituality, knowledge, lands and resources, and to their related practice of sustainable livelihoods."

Earth Charter principles 8 through 12, which bridge Sections II, III, and IV, clearly illuminate the reciprocal relation between the economic and environmental dimensions of human rights. They effectively reassert the substance of Environmental Justice Principles such as were first articulated in

* The Earth Charter offers a coherent, integrated standard for evaluating possible responses to particular environmental issues, immediate public policy choices, business and professional codes of conduct, and plans for change in community life.

the 1991 National People of Color Environmental Leadership Summit in Washington, D.C.

Democracy, Nonviolence, and Peace
Section IV, Democracy, Nonviolence, and Peace, emphasizes the importance of the fourth eco-justice norm: democratic participation in decisions about obtaining sustenance and managing the commons for the good of all. Principle 13 calls us to strengthen democratic institutions at all levels; to "provide transparency and accountability in governance, inclusive participation in decision making, and access to justice." Without these, there will be little healthy environmental, economic, and social policy. Principle 14 points to a parallel agenda for education, the arts, and the media: "Integrate into formal education and life-long learning the knowledge, values, and skills needed for a sustainable way of life." Toward this end, educators in the natural and social sciences as well as the humanities are utilizing the Earth Charter as a roadmap for ethically significant learning.

Principle 15, "Treat all living beings with respect and consideration," asserts humane concern for individual animals as a necessary complement to protecting ecosystems. Subprinciples focus on protecting sentient animals from suffering in confinement or through cruel methods of hunting, trapping, and fishing. The Earth Charter thus takes a step beyond previous international documents in stating an emerging, though still fragile, consensus about how humans should treat animals.

The Earth Charter's concluding principle 16—"Promote a culture of tolerance, nonviolence, and peace"—has special theological-ethical resonance: Its subprinciples concisely affirm peacemaking in the Gandhian tradition of active nonviolence, the Buddhist emphasis on nonharm, the Franciscan tradition of humble, peaceful living with neighbors and nature, and the peace church traditions that demand demilitarized societies for the sake of all life. In congruence with the biblical term *shalom* (peace with justice), the last supporting principle recognizes that "peace is the wholeness created by right relationships with oneself, other persons, other cultures, other life, Earth, and the larger whole of which all are a part." This is a good example of language in the Charter that is designed to resonate with people of diverse religious and cultural backgrounds. Nontheists are comfortable with a reference to "the larger whole of which all are a part," while monotheists can interpret that phrase to affirm the presence and power of God as the Sacred Whole, the source, sustainer, and reconciler of all being.

The Charter's four sets of ethical principles, it turns out, are similar to four key values of eco-justice ethics emphasized in ecumenical Christian thought over the last quarter century. Ecumenical leaders discerned in the mid-1970s that there would be little environmental health without social justice, and vice versa. Hence the ecumenical movement articulated principles for a global ethic featuring four norms that are both ends-oriented and means-clarifying:

- Solidarity with other people and creatures —companions, victims and allies—in Earth community, reflecting deep respect for creation;
- Ecological sustainability—environmentally fitting habits of living and

working that enable all life to flourish, utilizing ecologically and social appropriate technology;
- Sufficiency as a standard of organized sharing, which requires basic floors and definite ceilings for equitable or 'fair' consumption;
- Socially just participation in decisions about how to obtain sustenance and how to manage community life for the good in common and the good of the commons.

Likewise, Earth Charter principles 1 through 4 focus on solidarity, which comprehends the full dimensions of human-Earth relations and of inter-human obligation. Ecological sustainability is the focus of principles 5 though 8. Sufficiency—enough for all, justly distributed—a crucial criterion of socio-economic justice and human environmental rights, is illumined in principles 9 through 12. And democratic participation, animated by a culture of peace, is the concern of principles 13 through 16.

The Earth Charter's Importance for Biodiversity

The Earth Charter helps people of all ages in every walk of life to recognize global/local patterns of eco-injustice and unsustainable living, and to catch the spirit and substance of truly sustainable development that respects all life. In addition to showing what sustainable living means personally and collectively, the Earth Charter offers a coherent, integrated standard for evaluating possible responses to particular environmental issues, immediate public policy choices, business and professional codes of conduct, and plans for change in community life. In other words, it is quite useful as an ethical frame of reference for evaluating community and institutional practices and choosing among policy options. Readers of this essay should take note that all four parts of the Charter offer ethical principles and sub-principles for biodiversity conservation. Only through wide observance of these guidelines will biodiversity be sustained.

Our shared vocation is to care for disregarded places, species, and people, while reconciling humankind with the natural world. As the fourth paragraph of the Preamble declares, "The Choice is ours: form a global partnership to care for Earth and one another, or risk the destruction of ourselves and the diversity of life. [Toward this end,] fundamental changes are needed in our values, institutions, and ways of living."

To read the complete *Earth Charter*, see Appendix II at the end of the handbook.

Notes

[1] Bill McKibben, the author of *The End of Nature* (1989), coined this term in an article entitled "Climate Change and the Unraveling of Creation" in a December 8, 1999, issue of *The Christian Century*, where he wrote, "we are engaged in the swift and systematic decreation of the planet we were born onto . . . "

[2] Steven C. Rockefeller, "Global Interdependence, the Earth Charter, and Christian Faith," in Dieter Hessel & Larry Rasmussen, eds., *Earth Habitat: Eco-Injustice and the Church's Response* (Minneapolis, MN: Fortress Press, 2001). Rockefeller presented his paper to an October 1998 conference at the Union Theological Seminary on "Ecumenical Earth: New Dimensions of Church and Community in Creation."

[3] Thomas Berry, author of *The Dream of the Earth* (1988) and *The Great Work: Our Way Into the Future* (1999) coined this phrase in an article entitled "Community in Nature", in the Community issue of *The Living Pulpit* 3, no. 4, (October-December 1994), 28.

[4] See the words of Jesus in *John* 10:10.

Mutts

© Patrick McDonnell. Reprinted with special Permission of King Features Syndicate.

SECTION V

Communications Tips and Tools: Talking About Biodiversity, Ethics, and Faith

The Art of Communicating about Ethics

KEY POINTS

Communicating about ethics requires a thoughtful approach and an understanding of your audience's values.

Partnering with the religious community requires sensitivity and commitment.

Values-based messages should address your audience's values and concerns, explain the problem, and provide a solution.

✳ *The art of communicating effectively about ethics and moral choices is to avoid a debate on personal values, and instead find the common ground that leads to a wise course of action.*

by Jane Elder

The ethical argument is powerful but requires a thoughtful and sensitive approach.

Ethics, morals, and values are very personal and deeply held. They can charge a public debate like little else, because they articulate what is most important to us and why. The art of communicating effectively about ethics and moral choices involves avoiding debate on personal values, and instead finding the common ground that leads to a wise course of action. Here is some general guidance.

Tips and Reminders

- *Most people already have a strong sense of what is right and wrong—a firm set of values.* Communication with adults is not about persuading them to have values, or to get the right values, or worse yet, accusing them of not having any values. We may not like someone else's values, but we aren't going to get very far trying to change them. Values are shaped by our cultures, our families, our peer groups, and our experiences in life. But, even in our diverse culture, we have many common values.

- *Values-based communication is not about conversion to a particular point of view.* Our job is not to persuade an individual or an audience to adopt exactly the same values and viewpoint that we do, but rather to understand what drives *their* sense of right and wrong, and to be able to frame our messages about biodiversity conservation in the context of the ethics and values that speak to them. It means connecting the dots between the values people already have and the concerns they have (or might have) about biodiversity issues, and then offering positive choices that people can act upon to address those concerns in a way that is consistent with their existing values. In order to understand what their values are, the communication needs to be a dialogue, not a diatribe. (For more tips on developing and using effective values-based messages, see the following essay, "Crafting and Using Values-Based Messages").

- *One message or one ethical argument rarely fits all people.* While people across the U.S. tend to have similar primary values, because they have different backgrounds and interests, they will apply those values in very different ways. This is what makes democracy complex and fascinating, and why messages pitched to "the general public" often fail to reach anyone in particular. This is why marketers segment their audiences into groups with similar characteristics or attitudes.

- *Our messages are competing in a world cluttered with pitches to the same values.* Because it is effective, marketers and advertisers are promoting everything from politicians to beer with values. For example, Chrysler has an ad campaign in which the tag line is "Drive = Love," and Chevy sold its Blazer with the slogan, "Security in an insecure world." Likewise, politicians offer "A better choice for your future" or "Responsible leadership, for a change." Americans live in a message-saturated culture, and many of those messages are driven by values.

To get through the clutter, we need to take advantage of time-tested techniques for communicating about ethics and moral choices. These techniques include finding and telling *compelling human anecdotes that illustrate how real people make ethical choices for the environment*, and stories with which our audience can identify. Another effective technique is using the modern equivalent of the parable—another kind of story that uses metaphors and illustrative examples to *help people connect the dots* in ways they might not have seen before.

Keep in Mind . . .

1. Avoid universal declarations, such as "All Christians believe X" or "any ethical person would do Y." Pronouncements on how other people think and believe and how they should act are invitations to be challenged (at a minimum) and hoisted on your own petard. *Instead*...Consider statements such as," Within the Christian tradition, there are many who point out that God's first commandment was to tend the garden—to care for Creation" or "throughout human history, most cultures have valued protecting and sustaining the natural world that sustains them." But make sure you've done your homework on the assertions first, and that you can identify which cultures you are talking about as well as the basis for your argument.

2. Avoid playing "my values are more righteous than your values" (especially in public communications). They might be (more righteous), but no one wants to hear about it. Few things are more tiresome than the self-righteous environmentalist. "Holier than thou" has never been a popular communications tactic, and there's no reason to think that environmental advocates will have any greater success with it than sanctimonious voices of the past. *Instead*...Artfully listen for the values that offer potential for comment or question,

Social science and public opinion research have found that the following values are those most widely and deeply held across the U.S.

Primary American Values
- **Responsibility to care for one's family**
- **Responsibility to care for oneself**
- **Personal liberty**
- **Work**
- **Spirituality**
- **Honesty/integrity**
- **Fairness/equality**

In addition to these, there are other widely held values that are important, but not as important as the primary values.

Secondary Values
- **Responsibility to care for others**
- **Personal fulfillment**
- **Respect for authority**
- **Love of country or culture**

Of the broad range of values in American culture (including those above), the following are most commonly linked to environmental concerns:
- **Responsibility to care for the Earth and future generations**
- **Responsibility to one's family**
- **Responsibility to oneself**
- **Spirituality and sacredness of nature, respect for God's creation**
- **Personal fulfillment—enjoyment and aesthetics**
- **Love of country or culture**
- **Personal liberty and fairness**

Of these, those most strongly associated with the need to protect biodiversity are:
- **Responsibility to care for the Earth and future generations** (sometimes referred to as the stewardship value, although this term is not widely used in American conversation).
- **Respect for God's creation**

Another value relevant to the protection of biodiversity is responsibility to one's family (particularly as it relates to making sure your family enjoys a healthy, functioning environment). Also relevant, but less salient, are appreciation for the beauty of nature, national heritage, and the intrinsic value of nature.

Biodiversity Project Ethics for a Small Planet: A Communications Handbook 93

and then re-direct the argument: "But, Mr. X, you're a parent, too—surely you're not arguing that the kind of world our children will inherit doesn't matter."

3. Don't debate scripture, chapter and verse, tit for tat. This no-win strategy is a version of "my interpretation is superior to your interpretation" or "I know as many Bible passages as you do." The thoughtful comeback to a skewed interpretation of a particular passage is one thing, but claiming to have the "right" interpretation of scripture is presumptive and sets you up for a 2,000 year-old fight that is unlikely to be settled by you. Leave this to the theologians.

4. Don't use the inclusive "we" and "our" when framing community and cultural values, if you are not an authentic part of the community or the culture. This just opens you up for "who are you calling *we*" challenges. *Instead* be more cautious, and consider statements such as, "I believe I speak for many people in the community, when I say that I think there's something just plain wrong about destroying a wetland that the Creator entrusted to us."

Overall, the goal is not to have a moral or ethical showdown in public communications, but rather to illustrate the ethical and moral dimensions of the debate. This places the topic at hand into a context of common values and adds a human dimension to the issue. Let's use the community wetlands example again. In testimony, or in talking to a group or a reporter, you might say something like this:

"Scientists tell us that these wetlands are valuable to our local ecosystem, and the ecosystem is important, but I think there's something larger at stake here. We have the opportunity to do what's right for our community and for the people who will live here for generations to come. It might be cheaper in the short run to fill in these wetlands for new construction, but imagine what it will cost us in terms of the loss of beauty in our community, a child's chance to hear the first bird of spring in our own neighborhood, the chance to just be still and witness God's Creation right here in our town everyday. Losing these things is too high a cost—these are things that money can't buy.

5. Don't "preach" if you aren't ordained. Americans expect religious viewpoints from religious leaders. They are much less comfortable when someone without religious credentials begins to make religious claims in public. Remember, when Interior Secretary Babbitt made public appearances during which he talked about God's Creation, there was always someone next to him with a collar and a credential to embellish the observation. Babbitt expressed the view; the minister affirmed the credibility of the viewpoint. It doesn't mean that your convictions can't be voiced, but they need to be stated in a context that doesn't alienate your audience. For example, when you express your beliefs, make sure to provide the context: "As someone who has been active in my church all my life, I am led by my faith to consider X when I look at issues like this." This explains your role and your reason and allows an audience to hear you as a deeply religious individual, not a self-appointed interpreter of scripture. It creates a big difference in how you are received and what people hear.

Because ethics and moral principles are closely linked to many people's religious beliefs, building partnerships with the faith community can provide an important bridge to people who are already deeply invested in an ethical tradition. This outreach requires a thoughtful and sensitive approach.

Tips for Outreach to the Faith Community

(Adapted from Suellen Lowry and Daniel Swartz, *Building Partnerships with the Faith Community: A Resource Guide for Environmental Groups* [Madison, WI: Biodiversity Project, 2001])

Forging Relationships with Leaders in the Religious Community

There are a number of places to find religious community partners, including the following:
- Within your own environmental organization;
- In the yellow pages of the phone book;
- On the Internet (A good place to start a web search is the Web of Creation website at www.webofcreation.org, or the National Religious Partnership for the Environment website at www.nrpe.org.);
- At denominational regional offices;
- In social justice and conservation groups within denominations;
- At colleges and universities affiliated with denominations;
- In interfaith and ecumenical groups.

The Approach

- Before making an initial call, step back and ask yourself, "What would I be thinking if I had never before considered doing anything pertaining to biodiversity issues?"
- Reach out to lay members of the religious community as well as clergy.
- Avoid strident-sounding tones.
- Do not "put down" your opponents.
- Make a connection with issues on which individuals already are working, e.g., showing the relationship between biodiversity and social justice.
- Emphasize the many reasons that biodiversity is important, including species' inherent value and biodiversity's importance to people.

Partnering to Reach the Media

- Be careful when encouraging religious conservationists to do media work. By their very nature, media activities are not private, and many people want to keep their religious beliefs and practice private.
- Ask religious community members who are communicating with policy makers or the media to speak only from their own areas of expertise.
- Do not tell religious community individuals what their religious community message should contain, but share information from your own areas of expertise that may be helpful as the spiritual message is crafted.
- Ask whether you can contribute an article to the community's newsletter. Almost all religious communities have publications, often at the regional or national levels, and these publications may accept articles.

Ten Things to Think About

1. THE "religious community"
Perhaps the most basic mistake in outreach to religious groups is the assumption that such groups are all the same, all agree with each other, or all have the ability to speak for each other. Religious communities are as diverse as any other communities—often more so. Approach each group as its own entity, recognizing that even churches from the same denomination in the same town may be strikingly different from one another.

2. Evolution/Creation
One aspect of the religious community's diversity is the variety of approaches to evolution. Many congregations and religious leaders fully accept evolution; for others, the very term is anathema. Many congregations talk about "caring for God's Creation"—but they may mean very different things by that phrase. Find out what is and is not acceptable for a given congregation—understanding at the same time that groups all across the evolution/creation

spectrum may be supportive of biodiversity, though for different reasons.

3. Interfaith Coalitions and New Age/Pagan Issues

Another aspect of the diversity of religious life is the broad spectrum of reactions to interfaith coalitions. Some communities—especially the Jewish community—prefer to work in interfaith coalitions. Others—especially Evangelical churches—typically prefer to work independently. It is important to respect these differences and to encourage participation that is appropriate for a given congregation or leader. Some congregations and institutions worry that environmental groups or interfaith coalitions around environmental issues might be associated with "New Age" or "pagan" religious practices; other congregations welcome dialogue with Earth-based traditions or new religions. If you are working with communities where this is a concern, groups like the Evangelical Environmental Network can supply you with materials that offer biblically based reasons for "caring for creation." You need to make sure not to pressure religious community members to work publicly with any individuals or groups that might seem to compromise their religious beliefs.

4. Pro-Life Concerns

Religious groups also hold a variety of positions on abortion, birth control, and other family planning issues. While discussions of population issues and their relationship to biodiversity should not be considered off-limits, such discussions are sensitive and probably should wait until trust has begun to build in your relationships with religious leaders. In some cases, religious leaders will want to address environmental concerns as part of a broader "pro-life" agenda, an argument that can be extremely persuasive. It is important, however, for such themes to be raised by religious leaders rather than by secular environmental groups.

5. Diversity within Environmental Groups

Many religious communities have long-standing commitments to diversity in their own leadership, reflecting the great diversity within their pews. Such groups are particularly sensitive to criticisms of the environmental movement as an elitist concern held by upper-class white males. Be sure to demonstrate the diversity of your leadership to religious leaders, and if your group and leaders are not diverse, you might think about asking for help from religious leaders to increase your diversity.

6. Tax Status/Church-State Issues

These issues, while by no means limited to environmental concerns, are often raised by religious leaders who are relatively new to social justice activism. Many denominations publish very clear guidelines on activism, tax status, and church/state concerns for their congregation. Though you should not give detailed legal advice to religious groups, you can point out that religious leaders speaking out on environmental issues in no way threatens their tax status; the only activity they must avoid is the endorsement of political candidates. Similarly, if religious leaders are concerned about crossing church/state boundaries, you can point out that, as long as they do not try to establish a current religious test for public office, speaking about how their religious values relate to current policy questions does not intrude on church/state separation.

7. The "Enemy"

Especially in the present political climate, it is easy to fall into habits of demonizing one's political opponents. It is important to remember, however, that in a given congregation, one may find business leaders as well as environmental leaders, property rights activists as well as biodiversity activists. If religious leaders feel that they can't speak about environmental issues without directly attacking members of their own congregation, they may avoid the subject altogether. You can help by giving religious leaders the tools they need to express concern about environmental issues without ignoring legitimate questions about the consequences of environmental regulations. In some cases, once trust has been established, you may find that religious leaders can help bring other sectors to the table ready to work with you. In other cases, a trusting relationship may enable a religious leader to actively condemn practices by congregants that do need condemning.

8. Baggage
Sometimes the environmentalists approaching religious groups bring more than one agenda with them. Occasionally, in addition to environmental outreach, they want to discuss their own religious issues—anything from nagging questions about God to memories of a mean religious schoolteacher. While at least some of these conversations might be appropriate once a trusting relationship is built, they often can block the building of that relationship if they become too prominent too early on, and especially when they take on negative tones, denouncing religion as patriarchy or the like. Make sure that the person making the initial outreach connection to religious groups approaches them with an open mind and with only one item on the agenda—building a working relationship on environmental concerns.

9. Coming on Strong
Religious leaders are very busy, and they may not have previously given much thought to environmental concerns, especially biodiversity. Don't overwhelm them, either with reams of background information or with huge demands on their time. Make sure your initial contacts are about a time-limited, specific project. Once you have successfully worked together, you can move on to more complicated aspects of biodiversity and/or more long-term, time-intensive projects.

10. Two-Way Relationships
No one likes to be used, especially someone who already may feel overly busy and has too many demands on her/his time. Make sure the tone is not, "you'd be useful to me," but "maybe we can partner on this to accomplish something." Show clearly what you have to offer in this relationship—not only solid information about important subjects, but also exciting possibilities to make a difference on critical issues. And you can relate anecdotes about how environmental activities have brought new energy, excitement, and people, especially young people, to other congregations engaged in caring for creation. Finally, don't underestimate the value of helping provide interesting topics for sermons or religious school classes. The more you can offer, the more likely it is that religious leaders and groups will want to work closely with you.

ETHICS IN THE MEDIA

Movement Connects the Heavens with Earth

by Todd Wilkinson
The Oregonian, Portland, OR
December 26, 1999

On a crisp, wintry night, the Rev. Warren Murphy led parishioners on a walk beneath the glowing solstice moon. Together, they admired the cosmos with a telescope and sipped hot chocolate, and when the time came for a festive holiday carol, they sang "O Tannenbaum" with the enthusiasm of tree huggers.

The stroll is just the latest example of how Murphy, an Episcopalian priest, is persuading his flock to think "green" by paying regular tribute to the beauty of God's creation.

It's also part of a growing global movement involving spiritual leaders from all faiths asserting a strong connection among a healthy environment, spiritual fulfillment, and fundamental religious teachings.

From the Bible to the Talmud to the Koran, from weekend sermons to Christian rock concerts, Earth stewardship is emerging as a powerful religious force in the modern age. It is a trend, theologians say, that not only holds profound implications for religious and public policy from Capitol Hill to the Vatican, but also offers insight into how Americans view their biblical charge to care for God's creation.

The evolving synergy of the environmental and religious movements was documented in a survey by researchers at Harvard University in Cambridge, Mass. It showed a threefold increase in the number of people worshipping at environmentally focused churches during the mid-1990s.

This growth can be attributed to the increased interest of two particular demographic groups, says the Rev. Peter Illyan, Northwest regional director of Target Earth, one of several prominent eco-religious organizations:

> Young people who are active in the outdoors but raised without any firm religious teachings.

> Aging baby boomers who left their churches as young adults, feeling they were no longer relevant. Many are coming back because of their connection to contemporary environmental issues and the outreach of evangelical services.

Men and women of the cloth are drawing worshippers from all segments of society.

Most prominently, Pope John Paul II has quietly cultivated a legacy as the first environmental pope. In 1979, he proclaimed St. Francis the patron saint of ecology, and he has implored Roman Catholics to reduce their level of resource consumption.

"The seriousness of ecological degradation lays bare the depth of man's moral crisis," the pontiff declared on New Year's Day 10 years ago.

Religion frequently has entered environmental debates in Washington, D.C., as well. James Watt, the Interior Secretary under President Ronald Reagan and a born-again Christian, characterized environmentalists as practicing pagan idolatry for worshipping nature at the expense of the financial welfare of humans.

Watt claimed that natural-resource development has a firm rooting in Scripture—that man should have "dominion" over the land. From that assertion sprang a private-property rights movement in the West and South allied with fundamentalist Christians.

Leaders of the "green" religion movement admit they were slow to counter such assertions as they grew during the 1980s and '90s. But a turning point came in 1996, when Republicans in Congress wanted to amend the Endangered Species Act.

Clergy representing a spectrum of mainstream denominations protested, comparing the struggle to preserve biological diversity to Noah readying his ark. Newt Gingrich, House Speaker at the time, eventually shelved efforts to weaken the wildlife-conservation law.

Not universal support

Still, the eco-religious movement has its detractors. In the battle for support from evangelical Christians, both sides are armed with Bible passages to reinforce their point of view, and both accuse the other of misinterpreting Scripture.

E. Calvin Beisner, who teaches interdisciplinary studies at Covenant College in Lookout Mountain, Ga., is one of the nation's foremost critics.

He says that many ecological threats are overblown and that left-leaning environmentalists are trying to co-opt mainstream religion to add legitimacy to their cause.

"They infer that nature is best when it is pristine, and they say that man has fallen into sin by wishing to develop the landscape," he says. "They seem to suggest that everything man does has been negative."

Beisner, a devout promoter of the free market, and others say natural-resource development—including logging, mining, livestock grazing, and commercial fishing—helps accomplish a universal religious imperative, which is aiding the poor by elevating their quality of life.

"The Bible does specify that we have to be good stewards," says Michael Barkey, a policy analyst with the Acton Institute, a pro-business religious think tank. "While it seems like a very simple principle, it has broad economic ramifications."

Efforts by religious groups to end logging, for example, violate the separation of church and state, Barkey says. And he accuses certain religious groups of blasphemy by promoting Deep Ecology, which places humans not above nature to exercise dominion, but as merely a part of the ecosystem.

A lawsuit about logging in Minnesota, for example, is exploring whether the U.S. Forest Service views trees as "sacred."

But supporters of the new church activism in conservation say they're just responding to the wishes of congregations, which are both liberal and conservative.

"Our adversaries try to diminish our standing by labeling us part of the fringe," says Ann Alexander, chairwoman of the Christian Environmental Council. "Even if that were true—and it's not—it still wouldn't matter because millions of people are responding to our message because it is relevant."

Thousands of scientists, religious academics, ministers, and worshippers see no contradiction between evolution and creationism. Rather, they see a conduit between the two that closely parallels the objectives of environmentalism.

Movement has broad base

This movement manifests itself on a number of fronts:

A five-year-old program called Rescue God's Creation annually brings 50 Christian college students to Washington, D.C., to learn about environmental issues. When they return home, they use their new political insight to educate communities and fellow students about pending legislation.

The Pennsylvania Council of Churches began an unprecedented interfaith campaign to counter global climate change, saying it did "violence to God's creation" and violated moral and religious principles of justice.

The Religious Campaign for Forest Conservation is rallying hundreds of churches to support President Clinton's proposal to protect more than 40 million acres of public forests.

An effort led by the Redwood Rabbis, an extension of the Coalition on the Environment and Jewish Life in the

Northwest, staged a mock trial of a powerful timber executive, accusing him of violating Jewish law by felling ancient redwoods.

"We don't see it as a greening of religion as much as a drawing out of the inherent care of creation that has always been a part of Christianity," says Fred Krueger from the Religious Campaign for Forest Conservation. "The fact is you can't have a healthy economy and a severely degraded life-support system."

As the new millennium begins, when environmental concerns have never been greater, Illyan asks: "If Jesus were to appear today, would he be more inclined to be a land developer or a conservationist?"

"Scripture doesn't warn about worshipping nature," he says, "but it does warn continually about worshipping material wealth."

Copyright Oregonian Publishing Company, Dec 26, 1999. Reprinted by permission of Todd Wilkinson.

Bozeman, Montana, writer Todd Wilkinson is a western correspondent to the Christian Science Monitor, *a contributor to magazines such as* Audubon *and* National Geographic Adventure, *and author of nine books, including the critically acclaimed* Science Under Siege: The Politicians' War on Nature and Truth.

Crafting and Using Values-Based Messages

by Jane Elder

The Biodiversity Project has promoted values-based communications as an effective way to reach out to people with a wide range of backgrounds and interests. "Values-based" doesn't mean "value-laden." It also doesn't mean that we're trying to impose a particular set of values on someone. These false assumptions leave many people wary of this powerful communications tool, and they can miss out on an opportunity to expand the dialogue for biodiversity beyond the "choir" of environmental leaders who already understand why it is important.

Creating values-based messages simply means framing messages in a way that they speak to values that people already have, such as responsibility to future generations or a sense of fairness and honesty. By starting with values, we can speak to what people think is truly important, and thus we're more likely to open up a conversation. This isn't to say that the complexities of issues and the facts surrounding them aren't important—of course they are—but people tend to weigh them against an invisible scale of deeply held beliefs mixed with concerns for daily life and the future. Values-based messages are an invitation to weigh an issue through one's personal values. If they ring true, then you've really communicated.

A typical values-based message starts with a careful analysis of the intended audience and an analysis of what *they* value. The message should state what the problem, the issue, or the opportunity is and also provide a recommended solution or course of action to address the problem. But the problem and the solution need to be framed through the relevant values and concerns of the audience.

Let's say we want to talk with suburban parents about the importance of protecting local wetland habitat from development. We could start the conversation with wetland hydrology and its benefits to the local water table, the richness of amphibian species, and other facts, OR we could start by thinking about what the parents' values might be. Most parents care a great deal about the kind of world their children will inherit, as well as about their children's health and well-being. How might a wetland message speak to these concerns? Here's an example:

> Like other parents who live in Smith Meadows, we want our children to grow up in a healthy and beautiful community. The wetlands in our area help keep our drinking water clean and provide our families with an opportunity to enjoy wildlife throughout the year. But developers are seeking an exemption to our wetland regulations to build a new housing tract and shopping mall. We have a responsibility to our community and to our children's future to protect these special areas. By protecting the wetlands, we're protecting the quality of life in Smith Meadows.

Then, you can mention the water table and the amphibians if you need to, but start with responsibility, family, and future generations, and address concerns about health and quality of life. Starting with the values places the issue in the realm of doing what is right for one's community, one's family, and the future. The data can inform the decision, but the issue is no longer limited to merely data, and whose scientist has a better wetland map; it has expanded to include what our audiences (and we, too) value, and the ethical dimension has been engaged.

KEY POINTS

Messages that appeal to values and address concerns are the building blocks of an effective communications strategy.

Effective values-based messages are targeted and appropriate for the setting.

* Values-based messages are an invitation to weigh an issue through one's personal values. If it rings true, then you've really communicated.

Some tips on using values-based messages:

- *Targeting.* Values-based messages work well for targeted public communications, where the audience and its values are known. Public opinion research is one of the tools that enable communicators to identify audiences and their values. While many values are widely held throughout U.S. culture, the more specific the audience, the more incisive the message can be by speaking to the specific values that are important to your audience. If your target is the general public, you don't have a target audience.
- *We/Our.* Values-based messages tend to be most effective when they are inclusive: "We value our children's future" instead of "You value your children's future." But use inclusive language only when it is authentic.
- *Credible Messengers.* The messenger is just as important as the message (that's where that authentic "we" comes into play). The messenger needs to be authentic, credible, and persuasive to the audience you seek to reach as well as appropriate for the message itself. A neighborhood mom might work well for the wetlands message in the example above; a grandfather might be effective in talking about our connections to the land across generations; a clergyman is likely to be more effective than an environmental executive director in talking about the theological rationales for biodiversity protection.
- *Consider the Setting and the Application.* If you are participating in a technical hearing or a scientific debate, then stick to the language and terms that will communicate there. Values-based communications aren't a universal communications solution, although they work well for framing a public debate and reaching out to new audiences.

Sensitivities. Today educators are debating the highly charged issue of whether values belong in public education, and while values education is an entirely different kettle of fish than the communications strategies we're talking about here, it is possible to confuse the two. As a result, environmental educators need to ensure that they and their colleagues are clear about the distinctions, so the benefits of values-based communications in educational settings don't get dismissed because of concerns about values *in* education. Scientific and educational institutions, especially public institutions and agencies, are extremely sensitive to criticism that they are imposing an agenda on their visitors or students. Well-crafted messages don't do this, but communicators and educators still to need to be thoughtful in the use of values-based messages and responsive to the needs of their institution. One approach is to frame questions that bring social and cultural values into play but allow readers to draw their own conclusions. Using our example above, an exhibit or brochure might ask, "How do wetlands affect your quality of life?" and offer a range of choices for the visitor to consider.

A Message Is a Paragraph, Not a Slogan

A message is a clear, compelling, and short paragraph that does four basic things:
- Gives your audience a reason to care about your issue by appealing to values.
- Describes a threat and suggests who is responsible for the problem.
- Provides a solution.
- Describes what action will help solve the problem.

Making sure your message speaks to the way people sort through problems can strengthen a message. Does it appeal to our emotions? Does it provide us with information? Does it offer a solution? Does it give us something to do so we can respond to the threat?

A slogan and a sound bite can be easily lifted from your message. A slogan might be: "Development is forever." A sound bite might be: "We must protect our families' quality of life. If Smith marsh is destroyed, it's gone forever." These are shorthand extracts from your message that can be useful as a tag line in communications, but they are not a substitute for a thoughtful, well-constructed message paragraph.

Don't let parking lots be the only open spaces we leave our children. *[FUTURE GENERATIONS]*

Uncontrolled development destroys nearly 1 million acres of parks, farms, and open space each year. Sprawl not only threatens our environment, our health, and our quality of life, it's turning our communities into strip malls, freeways, and parking lots.

Protect your choices. Fight overdevelopment. Fight sprawl.

Biodiversity
Life. Nature. You. Make the connection.

[COMMUNITY] *[HEALTH]* *[QUALITY OF LIFE]*

This advertisement is available from the Biodiversity Project for other organizations to use. You can insert a call to action and your organization's contact information. To find out how to obtain a CD-ROM with a total of nine advertisements visit www.biodiversityproject.org/ads.htm

Biodiversity Project *Ethics for a Small Planet: A Communications Handbook* 103

Busting Anti-Conservation Myths

by Michael Nelson

KEY POINT

Understanding the logic of anti-conservation myths can prepare you to counter the premises and conclusions of opposing arguments.

✻ Although anti-conservation viewpoints are seldom presented as formal and systematic arguments, they are often built on premises and inferences that can be deconstructed and challenged.

Introduction

Ethical claims of various sorts are often presented as arguments. Therefore, our responses to those claims are most effective if we address them as arguments. Biodiversity advocates are frequently confronted with a common and predictable set of arguments against taking action to prevent extinction. Although anti-conservation viewpoints are seldom presented as formal and systematic arguments, they are often built on premises and inferences that can be deconstructed and challenged. Here are some examples of common anti-conservation claims—the kind you might encounter on talk radio—and ways to refute them using logical arguments.

I. The Naturalness Argument

The myth:
Extinction, climate change, soil erosion, air and water pollution, and acid rain are "natural" and normal phenomena; therefore we should not be concerned about them.

The responses:
1. *Not everything that is natural is good:*
   - Hurricanes are natural
   - Cancer is natural
   - Volcanic eruptions are natural
   - Strychnine is natural

   We are rightly concerned to protect ourselves from the first three, and we would be fools to willingly swallow the fourth.

2. *While these phenomena may occur in nature, people have increased the rate, frequency, and scale of their effects:*
   - The human-caused rate/degree of extinction, climate change, soil erosion, air and water pollution, and acid rain is much greater than the natural background rate.
   - Occasional catastrophic natural events (meteor strikes and volcanic eruptions) have caused infrequent spikes in the scale of mass extinctions, caused acid rain, and affected Earth's climate. But these occurrences cannot be compared to the way one species—*homo sapiens*—is altering the global environment, causing what may be irreversible changes, and threatening all species—including our own—with extinction.

II. The Human Superiority Argument

The myth:
Humans are superior to nature. Thus we have the right to use—or abuse—it in any way we wish or need.

The responses:
1. There are many standards by which to judge superiority:
- Not everyone agrees that humans are superior. Other species display intelligence, emotion, sensitivity to pain, complex systems of communication, the use of tools and (some researchers have argued) self-consciousness. Nonetheless, all of these capacities have at one time been described as uniquely human attributes allegedly testifying to our superiority.
- Some would say that the idea of "superiority" is an outmoded argument that has no practical meaning, given our current understanding of the complex interactions of species within the web of life.

2. People must act responsibly toward the rest of creation:
- If you are basing your conclusion of human superiority on religious teachings (e.g., the Bible), we will not argue with you about your religious beliefs concerning human primacy. But in our view the scriptures also teach that humans have a responsibility to be wise and careful stewards of nature.

3. Superiority does not grant the moral right to use and abuse others.
- With superiority comes a responsibility to care for those less privileged.

III. The "Conservation is Anti-Progress, Anti-People, and Unpatriotic" Argument

The myth:
Conservation is anti-progress, or anti-people, or unpatriotic.

The response:
1. What do you mean by "anti-progress", or "anti-people", or "unpatriotic"?
- Anti-progress: If by "progress" you mean *growth*, then our answer is that growth does not necessarily equal progress, just as change does not always equal progress. We question growth that will cause damage to essential parts or functions of the ecosystem.
- If by "progress" you mean *economic development*, conservation is certainly not anti-economic. The long-term economic security of our country relies on the availability of resources. Conservationists are very concerned about this and advocate policies that promote a sustainable way of life and renewability of resources. True progress means that we can live on this planet in a sustainable way, with resources being renewed rather than depleted.
  - Anti-people: Conservationists are trying to ensure the long-term health of our planet—and that is definitely good for people.
  - Unpatriotic: Questioning the status quo or "the way things have always been done" is not seditious. Questioning the way things are is one way to make things better. It is what democracy and free speech are all about. Conservationists question and oppose policies that will degrade the biosphere's life support systems and endanger many species, including our own.

IV. The Scientific Uncertainty Argument

The myth:
Science can't provide definitive answers to many questions (e.g., Are we causing global warming?), because not all scientists agree on the answers. Therefore, (a) it is wrong to assume that we're taking the cor-

✻ Conservationists are trying to ensure the long-term health of our planet—and that is definitely good for people.

✻ I can make a difference as an individual, and even though I am an individual, I am not alone. There are many others who agree with me…and together we can make a big difference.

rect action, or (b) we should do nothing until we have a scientific consensus.

The response:
1. *Science rarely produces one hundred percent certainty.*
- Scientists are almost never in total agreement over particular issues or questions. Scientific knowledge advances by constant questioning and testing—that is the way science works. True science does not assume absolute knowledge or certainty.
- When there is a critical mass of agreement among scientists that something is probably true, then we would be foolish not to act based on that body of knowledge.
- Most of our actions are based on the probability of a particular outcome and not on an absolute certainty. If *nine* doctors say that you will die if you don't take a certain medication, and *one* says that she disagrees, the wiser course is probably to take the medication.

V. The Hopelessness Argument

The myth (version 1):
The environment is doomed, and we can't do anything about it, so there is no point trying.

The response:
1. *Although it's true that many species have become extinct, that doesn't mean the entire cosmos is doomed.*
- Human actions are causing the major threats to biodiversity. Because we are responsible for the problem, we can fix much of it. What is more, we have a moral obligation to try.

The myth (version 2):
Individuals cannot make a difference, so anyone who tries to make a difference is just wasting time and effort.

The response:
1. *People can make a difference, individually and collectively:*
- Dedicated individuals have made a huge difference throughout history. Look at Nelson Mandela, Mahatma Gandhi, and Mother Teresa—or someone who is making a difference in your community. Throughout history, change has occurred because of individuals.
- I can make a small difference as an individual, and moreover, I am not alone. There are many others who agree with me, and together we can make a big difference.

Talking about Biodiversity

by Jane Elder

A concept as majestic and complex as biodiversity is challenging to teach. Most Americans are not reminded of biodiversity by their daily routines. Yet they are aware of—and concerned about—species loss and understand that humans are largely responsible for causing it. To expand on this awareness and concern and translate it into action, we will need to educate people more fully about the concept of biodiversity and the importance of its conservation. We will also need to inspire people to believe that they can make a difference, that they know what to do and how to do it, and that they can succeed.

Define It
If you use the word biodiversity, explain what it means. Otherwise, talk about the web of life, nature, the natural world, ecosystems, habitats, etc.

Make It Real, Not Conceptual or Abstract
Talk about biodiversity in the context of real places, real ecosystems, real species, and real issues. Ground the abstract concept of "diversity of gene pools, species, and habitats" in real places and experiences. Illustrate with forests, river systems, deserts, coastlines, wetlands, etc. and the variety of life that depends on them, instead of statistics about global species loss.

Localize Whenever Possible; Emphasize Place
Use local examples and experiences—a real place or problem that people can identify with—to provide context and meaning, e.g., loss of local songbirds, loss of the region's sugar maple trees, destruction of a local marsh, invasions from zebra mussels or kudzu, etc. Eschew the exotic (Biodiversity: it's not just for rainforests anymore!) when the local example is available. As long as species loss is occurring in far away places, it remains an abstract concept.

Make the Human Connection: Health and Human Services
Thanks to nature, life itself is possible: Illustrate and explain how healthy ecosystems sustain human life, from fresh air and clean water to food, fiber, and fun.

- Healthy natural systems keep us healthy: Balanced ecosystems promote human health, from supplying clean water to protecting us from exotic viruses, exploding insect populations, and toxic pollution. Among Americans, health is the primary motivator for protecting the environment; fear of toxics is the #1 concern.

- Nature's pharmacy: The potential loss of future sources of medicines interests some audiences (e.g., younger adults) and not others. But don't just talk about medicines that might come someday from exotic places. Instead explain common medicines that have already come from nature (cortisone, for example, from South African plant roots, or digitalis, from foxgloves) to illustrate how important natural sources of medicines already are. Start with the familiar; bridge to the possible.

Find Common Ground with Common Values; Lead with Values, Follow with Facts
Most Americans believe that we have a responsibility to maintain a clean and healthy environment for our families and for the future generations that will inherit the world we leave behind. This sense of "stewardship" provides common ground for starting conversations, after which the facts can be introduced.

If the Value Fits, Use It
Not everyone looks at the natural world the same way. Some think we should protect it because it is the responsible thing to do for the next generation, others, because it is God's creation, others, because it is beautiful, others because they believe in the intrinsic value of nature, etc. Know which values your audience embraces before you invoke a particular value in your argument. When in doubt, retreat to stewardship.

Emphasize Responsibility and Opportunity; Offer Hope!
Explain how humans are responsible for loss of species and natural areas, but also explain how humans can help reverse this trend. There's nothing like the imminent collapse of planetary life support systems to really turn off an audience. Don't sugar-coat the bad news, but always offer hope,

alternatives, options: "there's another way of doing things."

Connect the Dots... Make the Relationships and Interdependence of Nature Clear
Talk about species or particular habitats in terms of relationships, and explain the links to human well being whenever possible. (E.g., we need spiders because they eat insects and thus keep the insect population in balance, which in turn protects humans from out-of-control insect populations.) People understand that nature is an interdependent system, but they don't know much about the specific relationships.

Take Advantage of a Basic Appreciation of the Balance of Nature to Expand Ecological Literacy
Most people appreciate the concept of nature as a balanced system, but many don't know what it takes for nature to stay balanced. Explain basic concepts, such as "diversity provides resilience/ lack of diversity makes systems vulnerable"; explain the value of predators, scavengers, and other "undesirable" species in terms of the whole system. Explain, explain, explain.

- Speak in Plain English (or plain Spanish, etc.)
- Avoid scientific, technical, and other jargon.

Resources

Belden & Russonello, R/S/M, Inc. *Human Values and Nature's Future: Americans' Attitudes on Biological Diversity, An Analysis of Findings from a National Survey.* Washington, D.C.: Belden & Russonello (for the Communications Consortium Media Center), October 1996.

Biodiversity Project. Briefings/Workshops/Working Groups 1996 and 1997: A Summary Report. Madison, WI: Biodiversity Project, 1998.

Getting the Ethical Message Out to the Public

by Jane Elder and Erin Oliver

How do we get the ethical message out to a broader audience? Media and community outreach are two basic tools that will spread the word. There are many excellent guidebooks on media outreach and community organizing, and we won't make an effort to condense all their strategies and tips here. Instead, we have compiled a brief summary of special considerations and opportunities linked to communicating about ethics. We encourage you to take advantage of resources such as the Biodiversity Project's communications handbook, *Life. Nature. The Public. Making the Connection,* the Jossey-Bass handbook, *Guide to Strategic Communications for Nonprofits* (published by the Communications Consortium Media Center), and the Sierra Club's *Grassroots Organizing Training Manual* for a deeper look at the general topics of media and community outreach.

Environmental Ethics and the News Media

Ethical considerations about biodiversity protection do have potential for coverage in the news media, but the things that make an item newsworthy (its news value) will shape how a story is framed. The news media look for …
- Controversy—is there a disagreement or a struggle; is there conflict between two sides; are there heroes or villains?
- Immediacy—is something important happening right now?
- Compelling human interest—is there something engaging?
- Novelty, surprise, or the unusual—is there something that will capture attention?
- Celebrity—is someone famous or powerful involved?
- Scandal— are we shocked?
- Proximity—is there a local connection?
- Catastrophe and tragedy—has something awful happened?
- Trend—is this the beginning of something new?

As a result, a news story can often polarize a debate by emphasizing disagreement instead of common ground, or by focusing on immediate events instead of long-term solutions. Personality may take priority over substance, or a story might highlight the oddity of environmental concerns being raised by religious leaders, for example, as opposed to shedding light on the concerns themselves. Knowing what the established practices in the news media are and how reporters, editors and producers tend to package and present stories enables a savvy communicator to avoid being boxed in or misrepresented in the news. In order to use the news media effectively:
- Determine when the news is the right place to pitch a story, and
- Anticipate the needs of the news media, and develop stories that match those needs when appropriate.

Often, the complexity and nuance in ethical perspectives aren't easily compressed into a sound bite or a headline or even the news section of the daily paper. (The exception to this is the editorial and opinion page.) Feature articles or series provide opportunities to add some depth and dimension that a short news story can't provide. For example, profiles on people who demonstrate ethical courage in our times have news value—both novelty and compelling human interest. But consider where you come across these types of stories. Often, they appear as feature articles

---

KEY POINTS

Be aware of what makes something "newsworthy" when talking to the media about environmental ethics.

Features and Op-ed pieces provides an ideal forum for reaching out to audiences on ethics.

Look for venues where you can develop an ongoing dialogue to include the ethical considerations of biodiversity.

∗ By closely observing the media, you can identify where ethical issues about the environment are getting covered.

in the "lifestyle" or "communities" section of a newspaper where the focus is more on how we live than what shocking thing happened today. Other places where stories on the ethics of biodiversity protection might appear are the religion section of newspapers, features, the regular columns focusing on "observations about life," the food and garden sections, holiday features (Thanksgiving is a natural), and even the business section. For environmental writers, the emerging trend of ethical and religious activism in environmental debates is another angle to pursue.

Local television news is more of a challenge, since every minute is designed to compete for ratings. But "Live at Five" or "Good Morning" type programs are often crying out for content and are more accessible than the prime time news slots tend to be. An interview on one of these types of programs presents an opportunity to showcase compelling people and their stories, as long as the subject isn't too weighty. In radio, local NPR affiliates and community radio programs—those with talk shows or regular environmental features—present opportunities to develop stories with more depth than most radio news allows.

By closely observing the media, you can identify where ethical issues about the environment are getting covered, and you can cultivate relationships with writers, reporters, producers, and editors who have an interest in covering these stories. Below are additional suggestions for cultivating coverage in media.

More than News, and Alternative Media (adapted and revised from Dave Dempsey, *Life. Nature. The Public. Making the Connection* [Madison, WI: Biodiversity Project, 1999].)

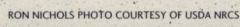

* Given the sensitivity associated with raising ethical and moral questions in public settings, look for venues where these larger questions may already be part of the dialogue.

- Local TV is increasingly entertainment-focused, even within news broadcasts. **Informational features** within regular news programming—"Spotlight on Health" or "the Weekly Garden Tip" segments—can be a good place to introduce biodiversity topics from a variety of angles, such as lifestyle concerns or consumer interests.
- Most newspapers have **weekly sections** on health and fitness, home and garden, religion, science, and other topics. For example, health section editors may be interested in a story about the way our increased dependence on the automobile can lead to health problems, and what this means for children growing up in an auto-dependent culture.
- Many cities have **alternative weekly newspapers,** such as the *Village Voice,* that cover issues of local interest. Many of these papers are free and thus are read by a large portion of urban dwellers. In addition to feature news articles, alternative weeklies often have sections offering brief updates on current issues.
- From the local to the national level, **magazines** are multiplying, providing new outlets for information. Commercial publications are increasingly aimed at "niche" audiences, ranging from outdoor recreationists to parents to community gardeners. Magazines know their audiences well and are a great tool for figuring out who, in our culture, is interested in what. Magazines also have "shelf-life" in a way that newspapers and electronic media don't. People often keep them and refer back to them, or they clip and save articles.
- **Electronic mail and websites** are now indispensable sources of information. (See the wealth of web sites on ethics, environment, theology, etc. in our reference section.) We can make good use

of electronic communications to expand the dialogue on biodiversity ethics, posting everything from bite-size factoids to in-depth analyses to tips on how to get involved in a local preservation effort.
- **"Talk radio"** has exploded in recent years. Although many of these shows have a conservative bent, you can use this forum to your advantage to counter stereotypes that the hosts and audience may bring to the discussion. Many public radio stations also host call-in or talk shows and are worth pursuing for coverage of local development issues.

Community Outreach (Adapted in part from recommendations in the Ecological Society of America's Communicating Ecosystem Services Toolkit. For a copy of the toolkit, contact ESA at www.esa.sdsc.edu/ecoservices)

Within any community there are people and organizations that care about the community's future and quality of life. Many of them may never have considered the role that biodiversity plays in shaping these things. Likewise, many communities have local environmental groups, nature centers, lake or park associations, etc., but these groups may not have explored the ethical issues at play in the community or in larger spheres. Given the sensitivity associated with raising ethical and moral questions in public settings, it is important to look for venues where these larger questions may already be part of the dialogue, such as:

- Community planning or "visioning" sessions that examine what we want our community to look like in 20 or 50 years;
- Interfaith groups that take on community problems or specific issues;
- Church study groups that explore a particular issue (such as environmental concerns) through their spiritual study;
- Community service projects or organizations that are designed to improve quality of life, clean-up rivers, protect important habitat, promote public transportation, etc. and,
- Neighborhood associations that are looking at quality of life issues, such as access to green space, community gardens, parks preservation, healthy lawns, etc.

Other opportunities to consider:
- Guest lecture in a college class.
- Look for relevant lecture series and offer to prepare a session or presentation.
- Build partnerships with interested leaders and congregations in the faith community (see the Biodiversity Project's *Building Partnerships with the Faith Community* for how to make contacts).
- Design a biodiversity service project that helps a local Boy/Girl Scout earn an environment or community service merit badge, or look for relevant projects with other youth organizations or 4-H clubs.
- Make a connection with horticulture and botany clubs, community garden groups, and plant/garden shops. Consider offering presentations, leading discussion groups, or creating exhibits/displays that educate about ways we value biodiversity.
- Connect with educational programs at museums, zoos, aquaria, botanical gardens, and nature and visitor centers.
- Scout the speaking opportunities and promote relevant service projects of chambers of commerce and service clubs.
- Reach out to outdoor recreation groups (fishing clubs, hiking clubs).
- Don't overlook the obvious—make the connection with local environmental organizations.

ETHICS IN THE MEDIA

Christians Tackle Climate Change Through International Conference

by Gordon Govier
The Capital Times, Madison, WI
July 9, 2002

A political football and a scientific dilemma will become a spiritual challenge at a conference in Oxford, England, this summer.

The Madison-based Au Sable Institute of Environmental Studies is teaming up with the United Kingdom's John Ray Initiative to sponsor an international forum on global climate change. It begins Sunday. The Au Sable Institute was founded by University of Wisconsin-Madison environmental studies professor Calvin DeWitt more than two decades ago, out of his belief in honoring God by honoring what God created.

In Sir John Houghton, chairman of the board of the John Ray Initiative, he has found a partner to help spread the gospel of environmental stewardship.

"We both are evangelical Christians who believe not only that we have to deal with the science of this, but that we also have to deal with the ethical implications for this, and for our work as Christians," says DeWitt.

Houghton is the former head of the United Kingdom's National Meteorology Office. That makes him one of the world's top weathermen. He's also co-chair of England's Scientific Assessment Working Group of the Intergovernmental Panel on Climate Change.

He and other scientists founded the John Ray Initiative three years ago, invoking the name of the pioneer botanist who first started classifying plants and animals three centuries ago. "We invited Cal DeWitt to come over and speak to us," he says, recalling the start of their collaboration. "One thing that seemed important to us was the whole idea of climate change, which many of us in the U.K.—but rather few in the U.S.—are very concerned about," Houghton says.

That begat this summer's conference.

Houghton hopes the leading scientists, economists, industrialists, theologians, and ethicists who convene will get the attention of people who have not been taking climate change seriously. "We want to look into ways to help," he says, "and get Christians to become better stewards of the environment and the resources we've been given."

"It's not immediately apparent what we should do" about global warming, says DeWitt. "We have to think through what our responsibilities are during these times. We have to deal with science as well as ethical implications."

> Stories that can be linked to dramatic events, such as natural disasters and catastrophes, may be more likely to appear in the news media.

The ethical implications that concern DeWitt include the impacts on populations, such as those that would be flooded out by rising ocean levels caused by global warming. Some South Pacific islands could be obliterated, and the Netherlands faces incredible expense to expand the dikes that hold back the ocean.

"This is all God's creation," DeWitt says. "It's richly beautiful. It's highly diverse, with a tremendous fabric of many kinds of living things. And whenever it is through our activities that this world is affected, we have to be concerned about our own behavior."

While the proofs of global warming are not always easy to pin down, DeWitt offers Lake Mendota as strong evidence. One of the most famous and most studied

> Stories that can be linked to an upcoming or recent event are more likely to get press coverage.

> Note the local angle, which adds relevance and appeal for local readers.

lakes in the world, where limnology actually began, it was recently projected to go a whole winter without freezing over by the year 2050.

But DeWitt says the lake failed to freeze this year, "at least as seen by satellite." He says one test indicated that only 60 percent of Lake Mendota froze.

That prompts him to observe that "this global warming trend is evident in the freeze and thaw dates for our lakes."

DeWitt's convictions are rooted deeply in a lifetime of passion for both scripture and nature. "We must not diminish this great second book which, along with God's first book—the Bible—are the two major means by which God makes himself known to us," he says.

To DeWitt, living an environmentally sensitive life is another way of worshipping God. He's encouraged by the thought that the teachings of the Gospels were not expounded in synagogues or temples, for the most part, but in God's great cathedral of the outdoors.

"In one sense Jesus almost always taught on field trips, like a biologist or ecologist would act today," he observes.

DeWitt realizes it will take more than one conference for people to embrace the concept of environmental stewardship. But he believes churches and denominations will inevitably have to address the issue.

"God's creation is not something we bow down to, but something we are entrusted with," he says. "The best way we can honor our creator is to honor the works of his hands, including ourselves."

Evangelicals not warm to ecology

by Gordon Govier

Many American Christians are wary of environmental issues like global warming because they associate environmentalism with New Age religious ideas like earth goddess worship, says recent University of Chicago divinity school graduate David Larson. Larson did his Ph.D. thesis on the Au Sable Institute.

He remembers well a conversation with his adviser, the distinguished Lutheran theologian Martin Marty, a decade ago.

"I had been raised as an evangelical. And in my entire life I had never heard of a sermon on an environmental theme," Larson recalls.

"I didn't know of any organizations that were promoting environmental issues from an evangelical perspective. He didn't know of any either but he said, 'why don't you look and see if you can find something?'"

Soon afterward, Larson stumbled across Au Sable Institute. But the uniqueness of its role was reinforced by other experiences.

"I went to a Christian bookstore at one point in the late '90s, when I was doing my research, and I asked if they had any books on the environment. The response of the attendant was, 'Do you realize you're in a Christian bookstore?'"

Although he has never met Calvin DeWitt, Larson has high praise for DeWitt's work through Au Sable Institute.

"They have been instrumental in crafting a theology that is distinctively evangelical, focusing on how Christians can better care for the environment."

Conflict: the media will almost invariably look for conflict—a basic device for holding the reader's interest.

←

The media are likely to look for personal stories that illustrate and humanize the article's subject. Personal stories help make an article about ideas more concrete, dramatic, and accessible. If you can, gather some personal stories in advance, before talking to the media, and have the people they focus unprepared to speak to reporters.

←

Reprinted by permission of Gordon Govier.

Seize the Days, Weeks, and Months:
A calendar of events and opportunities

In addition to community groups, keep in mind that making biodiversity real and localized whenever possible will help people make the connection between biodiversity and their own lives. Some events that may serve as opportunities to celebrate and educate are:

- Riverfront festivals
- Lake festivals
- Flower festivals
- Fruit Festivals
- Bird Counts
- Nature Walks
- Park Clean-ups
- Prairie Burns
- Farmers Markets

Holidays and commemorative days, weeks, or months are good opportunities to write a letter or sponsor a talk. Here's a quick overview of a year's worth of opportunities. Please note that some dates, such as Easter and Passover, change each year.

January
World Reverence for Life Awareness Month
Wild Bird Feeding Month
15th Day of Shvat, Tu B'shvat—Festival of Trees (Jewish)
January 1-7
Celebration of Life Week
January 1
Day of Meditation
January 1
National Environmental Policy Act anniversary
January 5
National Bird Day
January 6
Feast of Epiphany, Epiphany season begins and lasts until Ash Wednesday

February
February 1
Imbolc—Celebration to Welcome Spring (Pagan)
February 2
Groundhog Day
February 2
World Wetland Day
February 3
Endangered Species Act anniversary

March
Ethics Awareness Month
National Agriculture Week
Ash Wednesday-actual date depends on the date of Easter
March 21
Spring Equinox—Celebration of New Life
March 21
First Day of Spring
National Agriculture Day
March 21
First Day of Spring
National Flower Day
March 22
International Day of the Seal
March 22
United Nation's World Day for Water

April
Animal Cruelty Prevention Month
National Keep America Beautiful Month
National Lawn and Garden Month
World Habitat Awareness Month
National Garden Week
Grange Week
Week of Earth Day, National Park Week
Week of Earth Day, National Wildlife Week
Annually one week in April, National Week of the Ocean
Sixth week of the Easter season, Soil Stewardship Week
Last Friday in the month of April
National Arbor Day
April 13
Silent Spring publication anniversary
April 14
National Dolphin Day
April 21
John Muir's birthday
April 22
Earth Day
April 26
John James Audubon's birthday

May
Ascension Day, 40 days after Easter
Biodiversity Month
Clean Air Month
Flower Month
First full week in May, National Wildflower Week

Second week in May,
   National Historic Preservation
   Week
Second week in May,
   National Week of the Ocean
Third week in May,
   National Bike to Work Week
Seven weeks after Passover,
   Shavuot or Yom Habikkurim—
   Day of the First Fruits (Jewish)
May 1
May Day—Celebration of Flowers
May 1
Plant a Flower Day
May 1
Save the Rhino Day
Second Saturday in May
Migratory Bird Day
May 22
United Nation's International Day
for Biological Diversity
May 23
World Turtle Day
May 28
Whale Day

June
National Rivers Month
National Fishing Week,
   First week in June
National Garden Week,
   First Sunday in June
United Nation's Environmental
   Sabbath/Earth Rest Summer
June 1
National Trails Day
June 5
United Nation's World Environment
Day
June 8
World Oceans Day
June 20
National Bald Eagle Day
June 21
Summer Solstice—Celebration of
Growth
June 21
Midsummer's Night

July
National Parks and Recreation
Month

August
August 1
Lammas—Celebration of First
Harvest (Pagan)

September
National Organic Harvest Month
Third full week in September,
   National Farm Animals
   Awareness Week
September 19-23
Constitution Week
First Saturday after Labor Day
Federal Lands Clean Up Day
September 12
National Wildlife Ecology Day
September 16
United Nation's International Day
for Preservation of Ozone Layer
September 17
Citizenship Day
Third Saturday in September
International Coastal Clean-Up Day
September 21
Fall Equinox—Celebration of Harvest
September 28
National Public Lands Day
Fourth Saturday in September
National Hunting and Fishing Day

October
Week of the third Sunday in October,
   National Forest Products Week
Third week in October,
   National Wolf Awareness Week
October 21-29
World Rainforest Week
First Monday in October
United Nation's World Habitat Day
October 2
World Farm Animals Day
(also birthday of Mahatma Ghandi)
October 4
Feast of St. Francis—Blessing of the
Animals (Christian)

October 16
United Nation's World Food Day
October 18
Water Pollution Control Act/Clean
Water Act anniversary
October 21
Marine Mammal Protection Act
anniversary
October 31
Samhain—Celebration of Death and
Ancestors (Pagan)

November
Week ending with Thanksgiving,
   National Farm-City Week
November 1
Day of the Dead
November 2
World Ecology Day
November 13
World Kindness Day
November 15
America Recycles Day
Fourth Thursday in November,
Thanksgiving (U.S.)
November 19
National Community Education Day

December
End of November
through December 25,
Advent (Christian)
December 3
World Conservation Day
December 17
Clean Air Act anniversary
December 21
Winter Solstice
—Celebration of Rest and Renewal
December 21
World Peace Day
December 25
Christmas (Christian)
December 31
World Peace Meditation

ETHICS IN THE MEDIA

Editor's Note: Here's how Massachusetts' Executive Office of Environmental Affairs capitalized on Biodiversity Month to draw attention to biodiversity protection on Cape Cod.

Nature Needs Saving on South Shore

by Bob Durand
The Patriot Ledger, Quincy, MA
May 18, 2002

If you're heading down Route 3 toward the Cape, the signs of development are unmistakable. But what you might miss, if you are in a hurry, is the enormous variety of plant and animal life that inhabits what ecologists call the Southern New England Coastal Plains and Hills.

Although nearly half built out, the region is home to a surprising number of uncommon species and distinct natural communities. Black-crowned night herons nest out on the Harbor Islands, and the rare Blanding's turtle may be discovered in local bogs and wetlands.

May is national Biodiversity Month. This is the season to take stock in ourselves and the infinitely complex set of relationships that ensure our place on the planet. It is a time to recognize our reliance on this web of life. One of the greatest students of the natural world, Dr. Edward O. Wilson, calls that web "biodiversity." To draw on inspiration from Dr. Wilson's new book, "The Future of Life," the urgent quest to save Earth's flora and fauna, ourselves included, starts with ethics.

Here on the South Shore, the diversity of life is under constant assault. As the human footprint bears down on thousands of other living things with whom we share the region's ecosystems, we forfeit a circle of open space six miles in diameter—and a little bit of our future—every year. But that needn't be so.

In the past year, the commonwealth has taken strides to protect the state's rich biological heritage. The Executive Office of Environmental Affairs dedicated more than $50 million to open-space acquisition, and last summer met the goal of conserving more than 100,000 additional acres—three years ahead of schedule.

On the South Shore, the Neponset River Watershed Association has been managing Walpole's Willett Pond and its surroundings since acquiring this wetland not long ago.

NepRWA also advocated for the Metropolitan District Commission's purchase of the former Canton airport site, which, after remediation, will become another valuable addition to our precious open space. And the MDC itself is protecting nearly 5,000 acres in this most urban of watersheds, 816 of that in the past decade. Statewide, acting Gov. Jane Swift filed a comprehensive, $750 million environmental bond bill—the largest ever—that is now working its way through the Legislature. If enacted, the bond bill will help to ensure that these programs continue well into the future.

But we cannot move forward without a keen sense that preservation of habitat—and the commonwealth's more than 15,000 visible species—is the right thing to do. Without instilling a lasting environmental ethic in people, all the conservation programs in the world won't be enough to ensure that future generations will be able to enjoy nature's bounty.

Investment in our environment - and in the understanding that the wealth of life matters - is grounded in Biodiversity Days, a four- day festival of exploration and inquiry for all of the people of Massachusetts. This year's celebration is set for May 31-June 3, and nearly every city and town will participate. Most will offer programs in schools and field nature walks for both the curious among us and the dedicated expert.

In Quincy, for example, the Neponset River Watershed Association is partnering with the city's park department and the Environmental Treasurers of Quincy to organize a field trip on June 2 called "Life in the Salt Marsh." In Canton, Carl Lavin, a local naturalist and head of the Canton River Watershed Watchdogs, will lead a walk through open lands recently purchased by the town, accompanied by Canton High School's award-winning team of Problem Solvers. And in Milton, high school science teacher Barbara Plonski is planning walks on May 31 involving students in her biology class and local elementary schools. NepRWA also is working with the Blue Hills Observatory on a presentation the same day, "Measuring Biodiversity and the Weather." Still more events are planned.

This brings us to one of the great features of Biodiversity Days. You can decide what type of adventure you're up for, enter your requirements onto the web (see www.state.ma.us/envir/biodiversity.htm), and then generate a list of activities that suit your interests in your own community.

Biodiversity Days 2002 promises to be the largest event of its type ever held. Last year's celebration drew more than 30,000 participants. This year, we're expecting twice that many.

But our goal is not just numbers. It is a new environmental ethic. Species and ecosystems can't advocate for themselves. Biodiversity is protected only to the extent we want it to be. The state's biodiversity initiative is a critical step in preserving our natural systems for future generations. Without a broad constituency, we face the risk that this objective will be ignored until it's too late.

That's where the South Shore comes into the picture. And so do the citizens of Quincy and surrounding communities. Now is the time to explore the wonders of the natural world that we're all a part of. On Biodiversity Days, I hope you'll step outside and come take a walk with all of us.

Reprinted by permission of the Patriot Ledger. *Bob Durand is the Secretary of the Massachusetts Executive Office of Environmental Affairs.*

SECTION VI

Resources

© HENRY WOLCOTT 2001

GARY KRAMER PHOTO COURTESY OF USDA NRCS

Experts and Speakers on Biodiversity and Ethics

Peter Bakken
Coordinator of Outreach and Research Fellow
Outreach Office
Au Sable Institute
731 State Street
Madison, WI 53703
(608) 255-0950
bakken@ausable.org

Dee Boersma
Professor of Zoology
University of Washington
Department of Zoology
Seattle, WA 98195-1800
(206) 616-2185
boersma@u.washington.edu

Cassandra Carmichael
Director of Faith-Based Outreach
Center for a New American Dream
6930 Carroll Avenue, Suite 900
Takoma Park, MD 20912
(443) 822-3720
cassandra@toad.net

Anne Custer (speaker only – no media)
Research Associate
Forum on Religion and Ecology
P.O. 380875
Cambridge, MA 02238
(617) 244-6935
custer@fas.harvard.edu

Kathleen Dean Moore
Professor of Philosophy
Oregon State University
Department of Philosophy
Corvallis, OR 97331-3902
(541) 737-5652
KMoore@orst.edu

Scott Denman
Executive Director
Safe Energy Communication Council
1717 Massachusetts Avenue, NW, Suite 106
Washington, DC 20036
(202) 483-8491 x.*814
sdenman@erols.com

Cal DeWitt
Au Sable Institute
P.O. Box 260170
Madison, WI 53726
(608) 255-0950 Office
dewitt@ausable.org

Jane Elder
Executive Director
Biodiversity Project
214 North Henry Street, Suite 201
Madison, WI 53703
(608) 250-9876
jelder@biodiverse.org

Peter Forbes
Vice President & National Fellow
Trust for Public Land
Troubled Creek Farm
Canaan, NH 03741
(603) 523-7299
Peter.Forbes@TPL.org

Abby Kidder
Senior Education Associate
Institute for Global Ethics
11 Main Street
P.O. Box 563
Camden, ME 04843
(207) 236-6658
akidder@globalethics.org

Rhonda Kranz
Program Manager
Ecological Society of America
Sustainable Biosphere Initiative
1707 H Street, NW, Suite 400
Washington, DC 20006
(202) 833-8773 x.212
rhonda@esa.org

Suellen Lowry
Outreach Coordinator
Earth Justice Legal Defense Fund and
California Director
Interfaith Partnership for Children's Health
   and the Environment
1628 Hyland Street
Bayside, CA 95524
(707) 826-1948
suellenquaker@cox.net

Curt Meine
Wisconsin Academy of Sciences, Arts,
   and Letters
1922 University Avenue
Madison, WI 53705
(608) 263-1692 x.13
curt@savingcranes.org

Tom Muir
US Geological Service
MS 413 USGS
12201 Sunrise Valley Drive
Reston, VA 20192
tom_muir@usgs.gov

Gene Myers
Professor of Geography and
   Environmental Social Sciences
Western Washington University
Huxley College of Environmental Studies
Bellingham, WA 98225
(360) 650-4775
gmyers@cc.wwu.edu

Michael Nelson
Associate Professor of Philosophy
   and Natural Resources
University of Wisconsin-Stevens Point
Department of Philosophy
   and College of Natural Resources
Stevens Point, WI 54481
(715) 346-3907
Michael.Nelson@uwsp.edu

Bryan Norton
Professor of Philosophy
Georgia Institute of Technology
School of Public Policy

DM Smith Building, Room 300
685 Cherry Street
Atlanta, GA 30332-0345
(404) 894-6511
bryan.norton@pubpolicy.gatech.edu

Robert Perschel
Director
Land Ethic Program
The Wilderness Society
16 Germain Street
Worcester, MA 01602
(508) 756-4625
bob_perschel@tws.org

Edwin Pister
Executive Secretary
Desert Fishes Council
P.O. Box 337
Bishop, CA 93515
(619) 872-8751
phildesfish@telis.org

Daniel Swartz
Executive Director
Children's Environmental Health Network
110 Maryland Avenue, NE
Washington, DC 20002
(202) 543-4033 x.16
dswartz@cehn.org

Mitch Thomashow
Director, Doctoral Program in
Environmental Education
Antioch New England Graduate School
40 Avon Street
Keene, NH 03431
(603) 357-3122
mthomashow@top.monad.net

Mary Evelyn Tucker and John Grim
Professors of Religion
Forum on Religion and Ecology
Bucknell University
Department of Religion
Lewisburg, PA 17837
(570) 577-1205
mtucker@bucknell.edu
jgrim@bucknell.edu

Glossary of Terms

Anthropocentric: (1) Emphasizing the importance of humans in the scheme of things. (2) Regarding human interests as the only or the overriding consideration in ethics.

Anthropogenic: Originating in or caused by human activity.

Beliefs: Statements about facts or values that someone holds to be true, sometimes used to refer specifically to religious or moral teachings.

Biocentric: Regarding all forms of life as worthy of moral respect and consideration, and viewing human beings as one species among others.

Biophilia: Affiliation with and affection for the diversity of life forms, sometimes regarded as an evolved, innate characteristic of human beings.

Concerns: Matters regarded as immediately and practically relevant or important to a particular person or situation.

Constitutive: Essential to the very being of a thing; necessary for a thing to be what it is.

Cosmology: A theory or picture of the overall structure, composition, and contents of the universe, whether based on religious, philosophical, or scientific ideas, or some combination of these.

Eastern: Originating in or related to the cultures of Asia; Eastern Religions include Buddhism, Hinduism, Taoism, Shinto, Confucianism, and others.

Ecocentric: Ethically evaluating actions or policies in terms of their effects on ecological communities or ecosystems.

Ethics: (1) Rules defining how persons ought to behave toward one another or toward other beings (sometimes a synonym for morals), (2) The study of, or theories about, such rules.

Faith-based: Motivated by religious beliefs or values, or undertaken by a religious group or institution.

Holism: The belief that a complex whole has properties or values that cannot be reduced to the properties or values of its parts, and that an individual thing cannot be adequately valued or understood without reference to the whole(s) of which it is a part.

Instrumental: Being a means to an end; a thing has instrumental value when it is useful or beneficial for something else.

Intrinsic, inherent: Belonging to something independent of its relationship to other things; A thing has intrinsic or inherent value when it is valuable in and for itself.

Judeo-Christian: Originating in or related to the religious and ethical teachings and practices of Judaism and Christianity, particularly those that are common to the two religions.

Metaphysics: The study of, or a theory about, the nature, structure, and composition of reality that goes beyond what can be known by ordinary experience or scientific knowledge, or that analyzes such general and fundamental concepts as time, space, matter, causality, mind, etc.

Morals: Often a synonym for ethics, but may particularly connote customary beliefs about what constitutes proper behavior.

Ontology: The study of, or a theory about, what sorts of things there are and what characteristics they possess insofar as they exist, apart from their particular or individual natures; regarded either as synonymous with, or as a subfield of, metaphysics.

Precautionary principle: An ethical principle that states that if an action might harm human health or the environment, precautions should be taken, even if the likelihood of harm is scientifically unproven.

Religious: Relating to beliefs about and attitudes toward ultimate value, reality, and meaning, especially as articulated by a particular institution or tradition. Sometimes used to characterize anything that comprehensively orients a person's life and gives it meaning and value.

Rights: Moral or legal claims to be treated in a certain way, to be entitled to certain goods, or to be allowed to perform certain actions.

Spiritual: Involving a person's relationship to ultimate value, reality, and meaning; sometimes a synonym for "religious," but often used to contrast with institutional religious affiliations, outward practices, or doctrines.

Traditions: Teachings and practices (written or unwritten) that are handed down from generation to generation and, especially in the case of religious traditions, formative of communities which regard them as authoritative.

Utilitarian: (1) Having to do with a thing's utility or usefulness for a practical purpose, especially in contrast to spiritual, moral, or aesthetic value. (2) Relating to utilitarianism, an ethical theory in which an act is right if it leads to "the greatest good for the greatest number."

Values: (1) Things, qualities, conditions, etc. judged to be worthy of pursuing, protecting, or promoting as a general principle; (2) A person or group's attitudes about what is worthy or important.

Virtue: Qualities of character that enable or incline a person to act effectively, intelligently, or morally.

Worldview: A comprehensive understanding of reality and our place within it, which includes and integrates factual beliefs, moral values, and ultimate goals.

Bibliography (source material for this handbook)

Environmental Ethics and Values Bibliography

Attfield, Robin. *The Ethics of Environmental Concern*, 2nd edition. Athens, GA: University of Georgia Press, 1991.

Baxter, William. *People or Penguins: The Case for Optimal Pollution.* New York: Columbia University Press, 1974.

Beldon Russonello and Stewart. *Americans and Biodiversity: New Perspectives in 2002,* Madison, WI: The Biodiversity Project, 2002.

Callicott, J. Baird. *In Defense of the Land Ethic.* Albany, NY: State University of New York Press, 1989.

Callicott, J. Baird. *Beyond the Land Ethic.* Albany, NY: State University of New York Press, 1999.

Cooper, David A. and Joy Palmer. *Spirit of the Environment: Religion, Value, and Environmental Concern.* London, UK: Routledge, 1998.

Dempsey, Dave. *Life. Nature. The Public. Making the Connection: A Biodiversity Communications Handbook.* Madison, WI: The Biodiversity Project, 1999.

Devall, Bill, and George Session. *Deep Ecology. Living as if Nature Mattered.* Salt Lake City, UT: Peregrine Smith Books, 1985.

Forbes, Peter, Ann Armbrecht Forbes, and Helen Whybrow, eds. *Our Land, Ourselves: Readings on People and Place.* San Francisco, CA: The Trust for Public Land, 1999.

Fox, Warwick. "What Does the Recognition of Intrinsic Value Entail?" *The Trumpeter* 10, no. 3 (1993).

Fox, Warwick. *Toward a Transpersonal Ecology*, 2nd edition. Albany, NY: State University of New York Press, 1995.

Goodpaster, Kenneth. "On Being Morally Considerable." *Journal of Philosophy*, 75 (1978), 308-25.

Guha, Ramachandra. "Radical American Environmentalism and Wilderness Preservation: A Third World Critique." *Environmental Ethics* 11, no. 1 (1989), 71-83.

Hamilton, Lawrence S., ed. *Ethics, Religion, and Biodiversity: Relations Between Conservation and Cultural Values.* Cambridge, UK: White Horse Press, 1993.

Johnson, Lawrence. *A Morally Deep World: An Essay on Moral Significance and Environmental Ethics.* Cambridge, UK: Cambridge University Press, 1991.

Kellert, Stephen. *Kinship to Mastery: Biophilia in Human Evolution and Development.* Washington, DC: Island Press, 1997.

Kellert, Stephen, and Timothy Farnham, eds. *The Good in Nature and Humanity: Connecting Science, Spirit, and the Natural World.* Washington, DC: Island Press, 2002.

Kempton, Willett, James S. Boster, and Jennifer A. Hartley. *Environmental Values in American Culture.* Cambridge, MA: MIT Press, 1996.

Kidder, Abby. *How Big is Your Backyard?*

An Ethics-Based Approach to Environmental Decision Making. Camden, ME: Institute for Global Ethics, 2002.

Kidder, Rushworth M. *How Good People Make Tough Choices: Resolving the Dilemma of Ethical Living.* New York: Fireside Books, 1995.

Leopold, Aldo. *A Sand County Almanac: With Essays on Conservation from Round River.* New York, NY: Ballantine Books, 1966.

Mathews, Freya. *The Ecological Self.* London, UK: Routledge, 1991.

McKibben, Bill. *The End of Nature.* NY: Anchor Books, 1989.

Naess, Arne. *Ecology, Community, and Lifestyle.* Cambridge, UK: Cambridge University Press, 1989.

Nash, James A. "Biotic Rights and Human Ecological Responsibilities." *Annual of the Society of Christian Ethics*, (1993), 137-62.

Nash, Roderick. *The Rights of Nature: A History of Environmental Ethics.* Madison, WI: University of Wisconsin Press, 1989.

Norton, Bryan G. *Why Preserve Natural Variety?* Princeton, NJ: Princeton University Press, 1988.

Norton, Bryan G., ed. "The Moral Case for Saving Species." *Defenders*, 73, no. 3 (Summer 1998), 6-15.

Norton, Bryan G. "Biodiversity and Environmental Values: In Search of a Universal Earth Ethic." *Biodiversity and Conservation*, 9 (2000), 1029-44.

Palmer, Joy A., ed. *Fifty Key Thinkers on the Environment.* London, UK: Routledge, 2001.

Passmore, John. *Man's Responsibility for Nature*, 2nd edition. London, UK: Duckworth, 1980.

Perlman, Dan, and Glenn Adelson. *Biodiversity: Exploring Values and Priorities in Conservation.* Cambridge, MA: Blackwell Science, 1997.

Pister, Edwin Philip. "Species in a Bucket." Online Posting. *Natural History* (December 2000/January 2001), www.amnh.org/naturalhistory/editors_pick/1200_read5.html.

Plumwood, Val. *Feminism and the Mastery of Nature.* London, UK: Routledge, 1993.

Radford Ruether, Rosemary. *Gaia and God: An Ecofemenist Theology of Earth Healing.* San Francisco, CA: Harper, 1994.

Regan, Tom. *The Case for Animal Rights.* Berkeley, CA: University of California Press, 1983.

Rolston III, Holmes. "The Land Ethic at the Turn of the Millennium." *Biodiversity and Conservation* 9, no. 8 (2000), 1045-58.

Session, George, ed. *Deep Ecology for the Twenty-First Century: Readings on the Philosophy and Practice of the New Environmentalism.* Boston, MA: Shambhala Publishers, 1995.

Shiva, Vandana. *Staying Alive: Women, Ecology, and Development.* London, UK: Zed Books, 1989.

Shiva, Vandana. *Biodiversity: Social and Ecological Consequences.* London, UK: Zed Books, 1992.

Shiva, Vandana. *Tomorrow's Biodiversity (Prospects for Tomorrow).* New York, NY: Thames and Hudson, 2001.

Singer, Peter. *Animal Liberation*, 2nd edition. New York, NY: Random House, 1990.

Takacs, David. *The Idea of Biodiversity: Philosophies of Paradise*. Baltimore: The Johns Hopkins University Press, 1996, 194-287.

Taylor, Paul W. *Respect for Nature*. Princeton, NJ: Princeton University Press, 1986.

United Church of Christ, Commission on Racial Justice "Toxic Waste and Race in the United States: A National Report on the Racial and Socio-economic Characteristics of Communities with Hazardous Waste Sites." New York: United Church of Christ, 1987.

Warren, Karen, and Nisvan Erkal, eds. *Ecofeminism: Women, Culture, Nature*. Bloomington, IN: Indiana University Press, 1997.

Warren, Karen. "The Power and the Promise of Ecological Feminism." *Environmental Ethics*, 12, no. 2 (1990), 125-46.

Wenz, Peter. *Environmental Justice*. Albany, NY: State University of New York Press, 1988.

Wilson, Edward O. *Biophilia*. Cambridge, MA: Harvard University Press, 1984.

Wilson, Edward O. "Building an Ethic." Online Posting. *Defenders* (Spring 1993), www.defenders.org/bio-ee04.html.

Christian Theology and Religion

Anderson, Bernhard W. *From Creation to New Creation: Old Testament Perspectives*. Minneapolis, MN: Fortress Press, 1994.

Bakken, Peter, and Steven Bouma-Prediger, eds. *Evocations of Grace: Writings on Ecology, Theology, and Ethics*. Grand Rapids, MI: Eerdmans, 2000.

Ball, Jim. "Jesus Christ, Creation and the Protection of God's Creatures." *Green Cross: A Christian Environmental Quarterly* 2, no. 1 (1996), 8-12.

Berry, Thomas. *The Dream of the Earth*. San Francisco, CA: Sierra Club Books, 1988.

Berry, Thomas. *The Great Work: Our Way into the Future*. NY: Bell Tower, 1999.

Birch, Bruce C. *Let Justice Roll Down: The Old Testament, Ethics, and Christian Life*. Louisville, KY: Westminster/John Knox Press, 1991.

Bratton, Susan. *Christianity, Wilderness, and Wildlife: The Original Desert Solitaire*. Scranton, PA: University of Scranton Press, 1993.

Carroll, John E., and Keith Warner. OFM, eds. *Ecology and Religion: Scientists Speak*. IL: Franciscan Press, 1998.

Carroll, John E., Peter Brockelman, and Mary Westfall, eds. *The Greening of Faith: God, the Environment, and the Good Life*. NH: University Press of New Hampshire, 2000.

DeWitt, Calvin B. "The Price of Gopher Wood." *Faculty Dialogue* (Fall 1989).

DeWitt, Calvin B. "Biodiversity and the Bible." *Global Biodiversity* 6, no. 4 (1997), 13-16.

Dowd, Michael. *EarthSpirit: A Handbook for Nurturing an Ecological Christianity*. Mystic, CT: Twenty-third Publications, 1991.

Engel, J. Ronald. "Earth Spirituality is a Many Splendored Thing!" Online Posting. *The Journal of Liberal Religion* 1, no. 2 (Spring 2000), http://www.meadville.edu/jrengel_1_2.html.

Fowler, Robert Booth. *The Greening of Protestant Thought*. Chapel Hill, NC: University of North Carolina Press, 1995.

Fox, Matthew. *Creation Spirituality*. San Francisco, CA: HarperCollins, 1990.

Graber, Linda. *Wilderness as Sacred Space*. Washington, DC: Association of American Geographers, 1976.

Gustafson, James M. "Varieties of Moral Discourse: Prophetic, Narrative, Ethical and Policy." *Seeking Understanding: The Stob Lectures 1986-1998*, Calvin College and Calvin Theological Seminary, Grand Rapids, MI: Eerdmans, 2001.

Hall, Douglas John. *Imaging God: Dominion as Stewardship*. Grand Rapids, MI: Eerdmans, 1986.

Hessel, Dieter T. and Rosemary Radford Ruether, eds. *Christianity and Ecology: Seeking the Well-Being of Earth and Humans (Religions of the World and Ecology Series)*. Cambridge, MA: Harvard University Press, 2000.

Hessel, Dieter T., and Larry Rasmussen, eds. *Earth Habitat: Eco-Injustice and the Church's Response*. Minneapolis, MN: Fortress Press, 2001.

Hutchinson, Roger. *Prophets, Pastors, and Public Choices: Canadian Churches and the Mackenzie Valley Pipeline Debate*. Waterloo, Ontario: Wilfrid Laurier University Press, 1992.

LeQuire, Stan, ed. *The Best Preaching on Earth: Sermons on Caring for Creation*. Judson Press, 1996.

Lindner, Eileen, ed., and National Council of Churches of Christ. *Yearbook of American & Canadian Churches: Religious Pluralism in the New Millennium*. Nashville, TN: Abingdon Press, 2000.

Lovejoy, Arthur O. *The Great Chain of Being: A Study of the History of an Idea*. Cambridge, MA: Harvard University Press, 1936.

Lowry, Suellen, and Daniel Swartz. *Building Partnerships with the Faith Community: A Resource Guide for Environmental Groups*. Madison, WI: The Biodiversity Project, May 2001.

McDonagh, Sean. *The Greening of the Church*. Maryknoll, NY: Orbis Books, 1990.

McFague, Sallie. *The Body of God: An Ecological Theology*. Minneapolis, MN: Fortress Press, 1993.

Nash, James A. *Loving Nature: Ecological Integrity and Christian Responsibility*. Nashville, TN: Abingdon Press, 1991.

Oelschlaeger, Max. *Caring for Creation: An Ecumenical Approach to the Environmental Crisis*. New Haven, CT: Yale University Press, 1994.

Rasmussen, Larry. *Earth Community, Earth Ethics (Ecology and Justice Series)*. Maryknoll, NY: Orbis Books, 1997.

Santmire. H. Paul. *The Travail of Nature: The Ambiguous Promise of Christian Theology*. Minneapolis, MN: Fortress Press, 1985.

Sittler, Joseph. "A Theology for Earth." *Evocations of Grace*. Ed. Steven Bouma-Prediger and Peter Bakken. Grand Rapids, MI: Eerdmans, 2000.

Steck, Odil Hannes. *World and Environment*. Nashville, TN: Abingdon Press, 1980.

Stoll, Mark. *Protestantism, Capitalism, and Nature in America*. Albuquerque, NM: University of New Mexico Press, 1997.

United States Bishops'. *Renewing the Earth: An Invitation to Reflection and Action on Environment Ethics in Light of Catholic Social Teaching.* United States Catholic Conference, November 14, 1991.

Westermann, Claus. *The Living Psalms.* Grand Rapids, MI: Eerdmans, 1984.

White, Jr., Lynn, "The Historical Roots of Our Ecological Crisis." Science 155 (March 10, 1967), 1203-4.

Whitney, Elspeth. "Ecotheology and History." *Environmental Ethics* 15 (Summer 1993), 151-69.

Wilkinson, Loren, ed. *Earthkeeping in the 90's: Stewardship of Creation.* Grand Rapids, MI: Eerdmans, 1991.

Wolterstorff, Nicholas. *Until Justice and Peace Embrace.* Grand Rapids, MI: Eerdmans, 1983.

Worster, Donald. *The Wealth of Nature: Environmental History and the Ecological Imagination.* New York: Oxford University Press, 1993.

World Religions

Bassett, Libby, ed. *Earth and Faith: A Book of Reflection for Action.* NY: Interfaith Partnership for the Environment, United Nations Environment Programme, 2000.

Berkes, Fikret. "Religious Traditions and Biodiversity." *Encyclopedia of Biodiversity* 5 (2001), 109-20.

Dobb, Fred. "Overview on Jewish Texts on Biodiversity and Human Responsibility." Online Posting. *Coalition on Religion and Jewish Life,* www.coejl.org/learn/bd_source.shtml.

Golliher, Jeff. "Ethical, Moral, and Religious Concerns." *Cultural and Spiritual Values of Biodiversity.* Ed. Darrell Addison Posey et al. London, UK: Intermediate Technology Publications, United Nations Environment Programme, 1999, 435-502.

Gottlieb, Roger, ed. *This Sacred Earth: Religion, Nature, Environment.* NY: Routledge, 1996.

Kinsley, David. *Ecology and Religion: Ecological Spirituality in a Cross-Cultural Perspective.* Englewood Cliffs, NJ: Prentice Hall, 1995.

Macy, Joanna. *World as Lover, World as Self.* Berkeley, CA: Paralax Press, 1991.

Roberts, Elizabeth, and Elias Amadon, eds. *Earth Prayers from Around the World.* NY: HarperCollins, 1991.

Posey, Darrell Addison et al., eds. *Cultural and Spiritual Values of Biodiversity.* London, UK: Intermediate Technology Publications, United Nations Environment Programme, 1999.

Rockefeller, Steven, and John Elder, eds. *Spirit and Nature: Why the Environment Is a Religious Issue.* Boston, MA: Beacon Press, 1992.

Spring, David, and Ellen Spring. *Ecology and Religion in History.* NY: Harper and Row, 1974.

Swartz, Daniel, ed. *To Till and to Tend: A Guide to Jewish Environmental Study and Action.* NY: Coalition on the Environment and Jewish Life, 1993.

Tucker, Mary Evelyn, and John Grim, eds. *Worldviews and Ecology: Religion, Philosophy, and the Environment.* Maryknoll, NY: Orbis Press, 1994.

Communications

Bonk, Kathy and Henry Griggs and Emily Tynes. *The Jossey-Bass Guide to Strategic Communications for Nonprofits.* San Francisco, CA: Jossey-Bass Publishers, 1999.

Dempsey, David. *Life. Nature. The Public. Making the Connection: A Biodiversity Communications Handbook.* Madison, WI: The Biodiversity Project, 1999.

Sierra Club. *Grassroots Organizing Training Manual.* San Francisco, CA: The Sierra Club, 1999.

Additional Resources

Surveys and Bibliographies on the Environment, Ethics and Religion - Books

Bakken, Peter, Joan Gibb Engel and J. Ronald Engel. *Ecology, Justice, and Christian Faith: A Critical Guide to the Literature.* Westport, CT: Greenwood Press, 1995.

Davis, Donald Edward. *EcoPhilosophy: A Field Guide to the Literature.* San Pedro: R&E Miles, 1989.

Sheldon, Joseph K. *Rediscovery of Creation: A Bibliographical Study of the Church's Response to the Environmental Crisis.* ATLA Bibliography Series, No. 29. Metuchen, NJ: The Scarecrow Press, Inc., 1992.

Surveys and Bibliographies on the Environment, Ethics and Religion - Websites

Center for the Study of Values in Public Life
Harvard Divinity School
Subject Bibliographies in Environmental Ethics.
http://ecoethics.net/bib/

Earth Ministry
Greening Congregations Handbook: Stories, Ideas, and Resources for Cultivating Creation Awareness and Care in Your Congregation
www.earthministry.org

Forum on Religion and Ecology
http://environment.harvard.edu/religion/research/ethbiblio.html

International Society for Environmental Ethics
www.phil.unt.edu/ISEE/

Web of Creation
www.webofcreation.org/home2.html

Online Journals

Environmental Ethics
Center for Environmental Philosophy
P.O. Box 310980
University of North Texas
Denton, TX 76203-0980
(940) 565-2727
cep@unt.edu
www.cep.unt.edu/enethics.html

Environmental Values
The White Horse Press
10 High Street, Knapwell
Cambridge
United Kingdom CB3 8NR
44 1954 267527
www.erica.demon.co.uk/EV.html

Ethics and the Environment
Journals Division
Indiana University Press
601 North Morton Street
Bloomington, IN 47404
(812) 855-9449
journals@indiana.edu
www.iupjournals.org

International Society for Environmental Ethic
Max Oelschlaeger, Treasurer
Center for Community, Culture, and Environment
P.O. Box 5634
Northern Arizona University
Flagstaff, AZ 86011-5634
Max_Oelschlaeger@nau.edu
www.phil.unt.edu/ISEE/

Ecumenical, Interfaith, and Religious Organizations

Alliance of Religions and Conservation
3 Wynnstay Grove, Fallowfield
Manchester M14 6XG,
United Kingdom
arc_info@email.com
www.religionsandconservation.org

Coalition on the Environment and Jewish Life
433 Park Avenue South, 11th floor
New York, NY 10016-7322
(212) 684-6950 X. 210
coejl@aol.com
www.coejl.org

Earth Literacy Web
111 Fairmont Avenue
Oakland, CA 94611
(510) 595-5508
info@spirtualecology.org
www.spiritualecology.org

Earth Ministry
6512 23rd Avenue NW, Suite 317
Seattle, WA 98117-9923
(206) 632-2426
Emoffice@earthministry.org
www.earthministry.org

Interfaith Center for Corporate Responsibility
475 Riverside Drive, Room 566
New York, NY 10115
(212) 870-2295
www.iccr.org

National Council of Churches of Christ
Eco-Justice Working Group
475 Riverside Drive, Room 812
New York, NY 10115
(212) 870-2385
www.ncccusa.org or
www.webofcreation/NCC/workgrp.html

National Council of Churches of Christ
Environmental Justice Resources
P.O. Box 968
Elkhart, IN 46515
(800) 762-0968

National Council of Churches of Christ
Yearbook of American and Canadian Churches
475 Riverside Drive, Room 880
New York, NY 10115
(888) 870-3325
www.ncccusa.org

National Religious Partnership for the Environment
1047 Amsterdam Avenue
New York, NY 10025
(212) 316-7441
(800) 206-8858
nrpe@aol.com
www.nrpe.org

North American Coalition for Christianity and Ecology
Earthkeeping News
P.O. Box 40011
Saint Paul, MN 55104
(615) 698-0349
eudyson@worldnet.att.net
www.nacce.org

North American Coalition for Christianity and Ecology
Earthkeeping Circles Project
87 Stoll Road
Saugerties, NY 12477
(914) 246-0181
schaef@ulster.net

North American Coalition on Religion and Ecology
5 Thomas Circle, NW
Washington, DC 20005
(202) 462-2591
www.caringforcreation.net

Religious Campaign for Forest Protection
409 Mendocino Avenue, Suite A
Santa Rosa, CA 95401
(707) 573-3162
www.creationethics.org

Educational or Academic Organizations, Projects, & Programs

Alternatives for Simple Living
P.O. Box 2787
5312 Morningside Avenue
Sioux City, IA 51106
(712) 274-8875
(800) 821-6153
alternatives@SimpleLiving.org
www.simpleliving.org

American Academy of Religions
Religion and Ecology Group
825 Houston Mill Road, NE
Atlanta, GA 30329
(404) 727-3049
aar@aarweb.org
www.aarweb.org

Au Sable Institute
Outreach Office
731 State Street
Madison, WI 53703
(608) 255-0950
outreach@ausable.org
www.ausable.org

Center for Sacred Ecology
Earth Literacy Web
111 Fairmont Avenue
Oakland, CA 94611
(510) 595-5508
info@spiritualecology.org
www.spiritualecology.org

Forum on Religion and Ecology
Department of Religion
Bucknell University
Lewisburg, PA 17837
(570) 577-1205

Forum on Religion and Ecology
P.O. Box 380875
Cambridge, MA 02238
(617) 332-0337
fore@environment.harvard.edu
http://environment.harvard.edu/religion

Harvard Divinity School
Initiatives in Religion and Public Life
45 Francis Avenue
Cambridge, MA 02138
(617) 496-3586
irpl@hds.harvard.edu
www.hds.harvard.edu/irpl/

Harvard University Center for the Study of World Religions
Religions of the World and Ecology publications series
42 Francis Avenue
Cambridge, MA 02138
(617) 495-4486
www.hds.harvard.edu/cswr/ecology/index.htm
(Religions of the World and Ecology publications series also available through Harvard University Press, (800) 448-2242)

Harvard University's Pluralism Project
201 Vanserg Hall
25 Francis Avenue
Cambridge, MA 02138
(617) 496-2481
www.fas.harvard.edu/~pluralsm

Institute for Global Ethics
11 Main Street
P.O. Box 563
Camden, ME 04843
(207) 236-6658 or (800) 729-2615
www.globalethics.org

The Murie Center
P.O. Box 399
Moose, WY 83012
(307) 739-2246
muriecenter@wyoming.com
www.muriecenter.org

Theological Education to Meet the Environmental Challenge
2001 L Street, NW
Washington, DC 20037
(202) 778-6133
www.webofcreation.org/temecpage/temec

University of Creation Spirituality
2141 Broadway
Oakland, CA 94612
(510) 835-4827
www.creationspirituality.com

Whidbey Institute
P.O. Box 57
Clinton, WA 98236
(360) 341-1884
whidinst@whidbey.com
www.whidbeyinstitute.org

Environmental Organizations, Projects, and Programs

Center for a New American Dream
Faith-Based Outreach Program
6930 Carroll Avenue, Suite 900
Takoma Park, MD 20912
(301) 891-3683
newdream@newdream.org
www.newdream.org/faith

Center for Ethics and Toxics
P.O. Box 673
39120 Ocean Drive, Suite C-2-1
Gualala, CA 95445
(707) 884-1700
cetos@cetos.org
www.cetos.org

Center for Respect of Life and Environment
Earth Ethics newsletter (back issues)
2001 L Street, NW
Washington, DC 20037
(202) 778-6133
crle@aol.com
www.crle.org

Defenders of Wildlife
Defenders magazine
1101 14th Street, NW, #1400
Washington, DC 20005
(202) 682-9400
info@defenders.org
www.defenders.org

Earth Charter USA Campaign
2001 L Street, NW
Washington, DC 20037
(202) 778-6133
info@earthcharterusa.org
www.earthcharterusa.org

Earth Day Network
811 First Avenue, Suite 454
Seattle, WA 98104
(260) 876-2000
earthday@earthday.net
www.earthday.net

Forest Service Employees for Environmental Ethics
P. O. Box 11615
Eugene, OR 97440
(541) 484-2692
fseee@fseee.org
www.afseee.org

Institute for Ethics & Meaning
2109 Bayshore Boulevard., Suite 804
Tampa, FL 33606
(813) 254-8454 or (888) 538-7227
institute@transformworld.org
www.transformworld.org

National Association of Conservation Districts
Soil and Water Stewardship Week
P.O. Box 855
League City, TX 77574
(800) 825-5547, ext.28
www.nacdnet.org/pubatt/stewardship/index.htm

National Wildlife Federation
Backyard Habitat and NatureLink Programs
Education Outreach Department
8925 Leesburg Pike
Vienna, VA 22184
(703) 790-4483
www.nwf.org

Public Employees for Environmental Responsibility
2001 S Street, NW, Suite 570
Washington, DC 20009
(202) 265-7337
info@peer.org
www.peer.org

United Nations Environment Programme
Earth and Faith: A Book for Reflection and Action
DC2-803, United Nations
New York, NY 10017
(212) 963-8210
uneprona@un.org
www.unep.org

The Wilderness Society
1615 M Street, NW
Washington, DC 20036
(800) 843-9453
www.wildernesss.org

The Wilderness Society
Stories of the Land website
www.tws.org/ethic/stories.shtml

World Resources Institute
EcoStewards newsletter
409 Mendicino Avenue, Suite A
Santa Rosa, CA 95401
(707) 573-3160
wsi@ecostewards.org
www.ecostewards.org

World Wildlife Fund and Alliance of Religions and Conservation
Sacred Gifts for a Living Planet
www.panda.org/livingplanet

Publications

EarthLight: Magazine of Spiritual Ecology
Lauren de Boer, editor
111 Fairmount Avenue
Oakland, CA 94611
(510) 451-4926
www.earthlight.org

GeneWatch
Council for Responsible Genetics
5 Upland Road, Suite 3
Cambridge, MA
(617) 868-0870
www.gene-watch.org

Orion: People and Nature
195 Main Street
Great Barrington, MA 01230
(413) 528-4422 or (888) 909-6588
www.orionsociety.org

Orion Afield: Working for Nature and Community
195 Main Street
Great Barrington, MA 01230
(413) 528-4422
www.orionsociety.org

Science & Spirit: Connecting Science, Religion, and Life
Jennifer Derryberry, editor
115 Campbell Street, Suite L-4
Geneva, IL 60134
(630) 232-4002
info@science-spirit.org
www.science-spirit.org

Worldviews: Environment, Culture, Religion
Brill Academic Publishers
112 Water Street, Suite 400
Boston, MA 02109
(800) 962-4406

YES! The Magazine of Positive Futures
P.O. Box 10818
Bainbridge Island, WA 98110
(800) 937-4451
yes@futurenet.org
www.yesmagazine.org

Videos

Between Heaven and Earth: The Plight of the Chesapeake Watermen
Skunkfilms
(866) SKUNKFILMS
www.skunkfilms.com

Canticle to the Cosmos, with Brian Swimme
Center for the Story of the Universe
134 Colleen Street
Livermore, CA 94550
(800) 273-3720
www.brianswimme.org/centerspray.html

Catalogue of Video Resources on Ecology and Spirituality
Earth Communications
Lou Niznik
15726 Ashland Drive
Laurel, MD 20707
(301) 498-2553
www.rainforestjukebox.org/deep-eco/niznik.htm

Cherishing God's Creation
Presbyterian Distribution Service
100 Witherspoon Street
Louisville, KY 40202
(800) 524-2612

Earth Literacy Toolkit
Exploring a New Cosmology with Miriam Therese MacGillis, O.P.
Foundation for Global Community
Video Orders
222 High Street
Palo Alto, CA 94301-1097
(800) 707-7932
www.globalcommunity.org/cgvideo/

Faithful Earthkeeping: The Church as A Creation Awareness Center
Augsburg Fortress Publishers
Evangelical Lutheran Church in America
100 South Fifth Street, Suite 700
Minneapolis, MN 55402
(800) 328-4648
www.augsburgfortress.org

The Greening of Faith: Why the Environment is a Christian Concern
Cathedral Films & Video
P.O. Box 4029
Westlake Village, CA 91359
(800) 338-3456

Hope for a Renewed Earth
U.S. Conference of Catholic Bishops
3211 Fourth Street, NE
Washington, DC 20017
(800) 235-8722
www.nccbuscc.org/publishing/

Keeping the Earth, Religious and Scientific Perspectives on the Environment
Union of Concerned Scientists
Two Brattle Square
P.O. Box 9105
Cambridge, MA 02238
(617) 547-5552
www.uscusa.org

Love the Earth and Be Healed
EcuFilm
810 12th Avenue South
Nashville, TN 37203
(800) 251-4091
http://www.ecufilm.org

Other Relevant Websites

Adherents.com—Religious and Church Statistics
www.adherents.com

Alliance of Religions and Conservation
www.religionandconservation.org

American Religion Data Archive
www.thearda.com

Baptist Center for Ethics
www.Ethicsdaily.com

Beliefnet: The Source for Spirituality, Religion, and Morality
www.beliefnet.com

Earth Literacy Web
www.spiritualecology.org

Encyclopedia of Religion and Nature
www.ReligionandNature.com

Environmental Ethics, Center for Environmental Philosophy
www.cep.unt.edu

Ethics Updates, Environmental Ethics
http://ethics.acusd.edu/Applied/Environment/index.html

Forum on Religion and Ecology
http://environment.harvard.edu/religion

Gaia Nation
www.gaianation.org

Harvard Pluralism Project
www.fas.harvard.edu~pluralsm

Institute for Global Ethics
www.ige.org

Religion and Ethics Newsweekly
www.pbs.org/wnet/religionandethics

Web of Creation
www.webofcreation.org

Appendix I

The Assisi Declarations

Buddhism (1986)

There is a natural relationship between a cause and its resulting consequences in the physical world. In the life of the sentient beings too, including animals, there is a similar relationship of positive causes bringing about happiness, while undertakings generated through ignorance and negative attitude bring about suffering and misery. And this positive human attitude is, in the final analysis, rooted in genuine and unselfish compassion and loving kindness that seeks to bring about light and happiness for all sentient beings. Hence Buddhism is a religion of love, understanding and compassion, and committed towards the ideal of non-violence. As such, it also attaches great importance to wild life and the protection of the environment on which every being in this world depends for survival.

We regard our survival as an undeniable right. As co-inhabitants of this planet, other species too have this right of survival. And since human beings as well as other non-human sentient beings depend upon the environment as the ultimate source of life and well-being, let us share the conviction that the conservation of the environment, the restoration of the imbalance caused by our negligence in the past, be implemented with courage and determination.

Christianity (1986)

God declared everything to be good; indeed, very good. He created nothing unnecessarily and has omitted nothing that is necessary. Thus, even in the mutual opposition of the various elements of the universe, there exists a divinely willed harmony because creatures have received their mode of existence by the will of their Creator, whose purpose is that through their interdependence they should bring to perfection the beauty of the universe. It is the very nature of things considered in itself, without regard to humanity's convenience or inconvenience, that gives glory to the Creator.

Humanity's dominion cannot be understood as license to abuse, spoil, squander or destroy what God has made to manifest His glory. That dominion cannot be anything other than a stewardship in symbiosis with all creatures. On the one hand, humanity's position verges on a viceregal partnership with God; on the other, his self-mastery in symbiosis with creation must manifest the Lord's exclusive and absolute dominion over everything, over humanity and its stewardship. At the risk of destroying itself, humanity may not reduce to chaos or disorder, or, worse still, destroy God's bountiful treasures.

Every human act of irresponsibility towards creatures is an abomination. According to its gravity, it is an offence against that divine wisdom which sustains and gives purpose to the interdependent harmony of the universe.

Hinduism (1986)

Hinduism believes in the all-encompassing sovereignty of the divine, manifesting itself in a graded scale of evolution. The human race, though at the top of the evolutionary pyramid at present, is not seen as something apart from the earth and its multitudinous life-forms.

The Hindu viewpoint on nature is permeated by a reverence for life, and an awareness that the great forces of nature—the earth, the sky, the air, the water and fire —as well as various orders of life including plants and trees, forests and animals, are all bound to each other within the great rhythms of nature. The divine is not exterior to creation, but expresses itself through natural phenomena. In the Mudaka Upanishad the divine is described as follows:

'Fire is his head, his eyes are the moon and sun; the regions of space are his ears, his voice the revealed Veda; the wind is his breath, his heart is the entire universe; the earth is his footstool, truly he is the inner soul of all.'

The natural environment has received the close attention of the ancient Hindu scriptures. Forests and groves were considered sacred, and flowering trees received special reverence. just as various animals were associated with gods and goddesses, different trees and plants were also associated in the Hindu pantheon. The Mahabharata says that, 'even if there is only one tree full of flowers and fruits in a village, that place becomes worthy of worship and respect.'

Islam (1986)

The essence of Islamic teaching is that the entire universe is God's creation. Allah makes the waters flow upon the earth, upholds the heaven, makes the rain fall and keeps the boundaries between day and night. The whole of the rich and wonderful universe belongs to God, its maker. It is God who created the plants and the animals in their pairs and gave them the means to multiply. Then God created humanity a very special creation because humanity alone was created with reason and the power to think, and even the means to turn against the Creator. Humanity has the potential to acquire a status higher than that of the angels or sink lower than the lowliest of the beasts.

For the Muslim, humanity's role on earth is that of a 'khalifa', vice-regent or trustee of God. We are God's stewards and agents on Earth. We are not masters of this Earth, it does not belong to us to do what we wish. It belongs to God and He has entrusted us with its safekeeping. Our function as vice-regents, 'khalifa' of God, is only to oversee the trust. The 'khalifa' is answerable for his/her actions, for the way in which he/she uses or abuses the trust of God.

Allah is Unity; and His Unity is also reflected in the unity of humanity, and the unity of humanity and nature. His trustees are responsible for maintaining the unity of His creation, the integrity of the Earth, its flora and fauna, its wildlife and natural environment. Unity cannot be had by discord, by setting one need against another or letting one end predominate over another; it is maintained by balance and harmony.

Judaism (1986)

When God created the world, so the Bible tells us, He made order out of primal chaos. The sun, the moon, and the stars; plants, animals, and ultimately humanity, were each created with a rightful and necessary place in the universe. They were not to encroach on each other. 'Even the divine teaching, the Torah, which was revealed from on high, was given in a set measure' (Vayikra Rabbah 15:22) and even these holy words may not extend beyond their assigned limit.

The highest form of obedience to God's commandments is to do them not in mere acceptance but in the nature of union with Him. In such a joyous encounter between man and God, the very rightness of the world is affirmed. The encounter of God and man in nature is thus conceived in Judaism as a seamless web with man as the leader and custodian of the natural world.

There is a tension at the centre of the Biblical tradition, embedded in the very story of creation itself, over the question of power and stewardship. The world was created because God willed it, but why did He will it? Judaism has maintained, in all of its versions, that this world is the arena that God created for man, half beast and half angel, to prove that he could behave as a moral being. The Bible did not fail to demand even of God Himself that He be bound, as much as humanity, by the law of morality. Thus Abraham stood before God, after He announced that He was about to destroy the wicked city of Sodom, and Abraham demanded of God Himself that He produce moral justification for this act: 'Shall not the judge of all the earth do justice?' Comparably, man was given dominion over nature, but he was commanded to behave towards the rest of creation with justice and compassion. Humanity lives, always, in tension between his/her power and the limits set by conscience.

The following three statements, from the Baha'is, Sikhs and Jains, were issued in the years following Assisi.

Baha'i (1987)

'Nature in its essence is the embodiment of My Name, the Maker, the Creator. Its manifestations are diversified by varying causes, and in this diversity there are signs for men of discernment. Nature is God's Will and is its expression in and through the contingent world. It is a dispensation of Providence ordained by the Ordainer, the All-Wise.' (Baha'i writings.)

With those words, Baha'u'Uah, Prophet-founder of the Baha'i faith, outlines the essential relationship between humanity and the environment: that the grandeur and diversity of the natural world are purposeful reflections of the majesty and bounty of God. For Baha'is, there follows an implicit understanding that nature is to be respected and protected, a divine trust for which we are answerable.

As the most recent of God's revelations, however, the Baha'i teachings have a special relevance to present-day circumstances when the whole of nature is threatened by man-made perils ranging from the wholesale destruction of the world's rainforests to the final nightmare of nuclear annihilation.

A century ago, Baha'u'llah proclaimed that humanity has entered a new age. Promised by all the religious Messengers of the past, this new epoch will ultimately bring peace and enlightenment for humanity. To reach that point, however, humankind must first recognize its fundamental unity - as well as the unity of God and of religion. Until there is a general recognition of this wholeness and interdependence, humanity's problems will only worsen.

Sikhism (1989)

Since the beginning of the Sikh religion in the late fifteenth century, the faith has been built upon the message of the 'oneness of Creation'.

Sikhism believes an almighty God created the universe. He himself is the creator and the master of all forms in the universe, responsible for all modes of nature and all elements in the world.

Sikhism firmly believes God to be the source of the birth, life and death of all beings. God is the omniscient, the basic cause of the creation and the personal God of them all.

From the Divine command occurs the creation and the dissolution of the universe. (pll7 Guru Granth Sahib)

As their creator, the natural beauty which exists and can be found in all livings things whether animals, birds, fish, belongs to Him, and He alone is their master and without His Hukum (order) nothing exists, changes or develops.

Having brought the world into being, God sustains, nourishes and protects it. Nothing is overlooked. Even creatures in rocks and stones are well provided for. Birds who fly thousands of miles away leaving their young ones behind know that they would be sustained and taught to fend for themselves by God (Guru Arjan, in Rehras). The creatures of nature lead their lives under God's command and with God's grace.

Jainism (1991)

The Jain ecological philosophy is virtually synonymous with the principle of ahimsa (non-violence) which runs through the Jain tradition like a golden thread.

Ahimsa is a principle that Jains teach and practice not only towards human beings but towards all nature. It is an unequivocal teaching that is at once ancient and contemporary.

There is nothing so small and subtle as the atom nor any element so vast as space. Similarly, there is no human quality more subtle than non-violence and no virtue of spirit greater than reverence for life.

The teaching of *ahimsa* refers not only to physical acts of violence but also to violence in the hearts and minds of human beings, their lack of concern and compassion for their fellow humans and for the natural world. Ancient Jain texts explain that violence *(himsa)* is not defined by actual harm, for this may be unintentional. It is the intention to harm, the absence of compassion, that makes an action violent. Without violent thought there could be no violent actions.

Jain cosmology recognizes the fundamental natural phenomenon of symbiosis or mutual dependence. All aspects of nature belong together and are bound in a physical as well as a metaphysical relationship. Life is viewed as a gift of togetherness, accommodation and assistance in a universe teeming with interdependent constituents.

Appendix II

The Earth Charter

Preamble

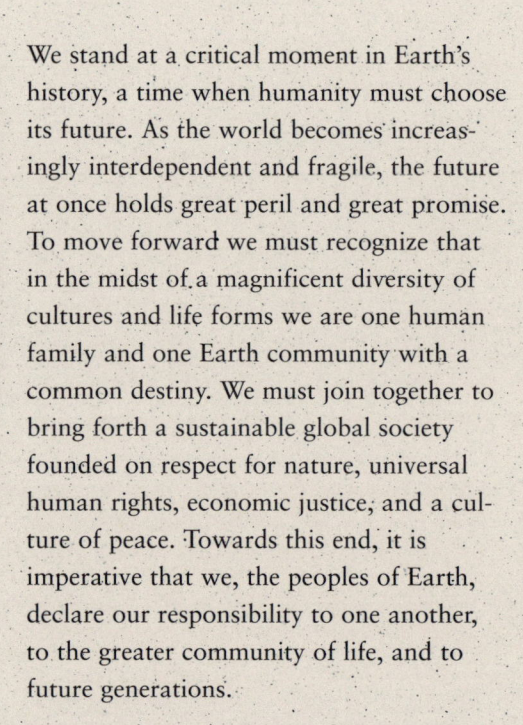

We stand at a critical moment in Earth's history, a time when humanity must choose its future. As the world becomes increasingly interdependent and fragile, the future at once holds great peril and great promise. To move forward we must recognize that in the midst of a magnificent diversity of cultures and life forms we are one human family and one Earth community with a common destiny. We must join together to bring forth a sustainable global society founded on respect for nature, universal human rights, economic justice, and a culture of peace. Towards this end, it is imperative that we, the peoples of Earth, declare our responsibility to one another, to the greater community of life, and to future generations.

Earth, Our Home

Humanity is part of a vast evolving universe. Earth, our home, is alive with a unique community of life. The forces of nature make existence a demanding and uncertain adventure, but Earth has provided the conditions essential to life's evolution. The resilience of the community of life and the well-being of humanity depend upon preserving a healthy biosphere with all its ecological systems, a rich variety of plants and animals, fertile soils, pure waters, and clean air. The global environment with its finite resources is a common concern of all peoples. The protection of Earth's vitality, diversity, and beauty is a sacred trust.

The Global Situation

The dominant patterns of production and consumption are causing environmental devastation, the depletion of resources, and a massive extinction of species. Communities are being undermined. The benefits of development are not shared equitably and the gap between rich and poor is widening. Injustice, poverty, ignorance, and violent conflict are widespread and the cause of great suffering. An unprecedented rise in human population has overburdened ecological and social systems. The foundations of global security are threatened. These trends are perilous—but not inevitable.

The Challenges Ahead

The choice is ours: form a global partnership to care for Earth and one another or risk the destruction of ourselves and the diversity of life. Fundamental changes are needed in our values, institutions, and ways of living. We must realize that when basic needs have been met, human development is primarily about being more, not having more. We have the knowledge and technology to provide for all and to reduce our impacts on the environment. The emergence of a global civil society is creating new opportunities to build a democratic and humane world. Our environmental, economic, political, social, and spiritual challenges are interconnected, and together we can forge inclusive solutions.

Universal Responsibility

To realize these aspirations, we must decide to live with a sense of universal responsibility, identifying ourselves with the whole Earth community as well as our local communities. We are at once citizens of different nations and of one world in which the local and global are linked. Everyone shares responsibility for the present and future well-being of the human family and the larger living world. The spirit of human solidarity and kinship with all life is strengthened when we live with reverence for the mystery of being, gratitude for the gift of life, and humility regarding the human place in nature.

We urgently need a shared vision of basic values to provide an ethical foundation for the emerging world community. Therefore, together in hope we affirm the following interdependent principles for a sustainable way of life as a common standard by which the conduct of all individuals, organizations, businesses, governments, and transnational institutions is to be guided and assessed.

Principles

I. Respect and Care for the Community of Life

1. Respect Earth and life in all its diversity.
 a. Recognize that all beings are interdependent and every form of life has value regardless of its worth to human beings.
 b. Affirm faith in the inherent dignity of all human beings and in the intellectual, artistic, ethical, and spiritual potential of humanity.

2. Care for the community of life with understanding, compassion, and love.
 a. Accept that with the right to own, manage, and use natural resources comes the duty to prevent environmental harm and to protect the rights of people.
 b. Affirm that with increased freedom, knowledge, and power comes increased responsibility to promote the common good.

3. Build democratic societies that are just, participatory, sustainable, and peaceful.
 a. Ensure that communities at all levels guarantee human rights and fundamental freedoms and provide everyone an opportunity to realize his or her full potential.
 b. Promote social and economic justice, enabling all to achieve a secure and meaningful livelihood that is ecologically responsible.

4. Secure Earth's bounty and beauty for present and future generations.
 a. Recognize that the freedom of action of each generation is qualified by the needs of future generations.
 b. Transmit to future generations values, traditions, and institutions that support the long-term flourishing of Earth's human and ecological communities.

In order to fulfill these four broad commitments, it is necessary to:

II. Ecological Integrity

5. Protect and restore the integrity of Earth's ecological systems, with special concern for biological diversity and the natural processes that sustain life.

 a. Adopt at all levels sustainable development plans and regulations that make environmental conservation and rehabilitation integral to all development initiatives.
 b. Establish and safeguard viable nature and biosphere reserves, including wild lands and marine areas, to protect Earth's life support systems, maintain biodiversity, and preserve our natural heritage.
 c. Promote the recovery of endangered species and ecosystems.
 d. Control and eradicate non-native or genetically modified organisms harmful to native species and the environment, and prevent introduction of such harmful organisms.
 e. Manage the use of renewable resources such as water, soil, forest products, and marine life in ways that do not exceed rates of regeneration and that protect the health of ecosystems.
 f. Manage the extraction and use of non-renewable resources such as minerals and fossil fuels in ways that minimize depletion and cause no serious environmental damage.

6. *Prevent harm as the best method of environmental protection and, when knowledge is limited, apply a precautionary approach.*
   a. Take action to avoid the possibility of serious or irreversible environmental harm even when scientific knowledge is incomplete or inconclusive.
   b. Place the burden of proof on those who argue that a proposed activity will not cause significant harm, and make the responsible parties liable for environmental harm.
   c. Ensure that decision-making addresses the cumulative, long-term, indirect, long distance, and global consequences of human activities.
   d. Prevent pollution of any part of the environment and allow no build-up of radioactive, toxic, or other hazardous substances.
   e. Avoid military activities damaging to the environment.

7. *Adopt patterns of production, consumption, and reproduction that safeguard Earth's regenerative capacities, human rights, and community well-being.*
   a. Reduce, reuse, and recycle the materials used in production and consumption systems, and ensure that residual waste can be assimilated by ecological systems.
   b. Act with restraint and efficiency when using energy, and rely increasingly on renewable energy sources such as solar and wind.
   c. Promote the development, adoption, and equitable transfer of environmentally sound technologies.
   d. Internalize the full environmental and social costs of goods and services in the selling price, and enable consumers to identify products that meet the highest social and environmental standards.
   e. Ensure universal access to health care that fosters reproductive health and responsible reproduction.
   f. Adopt lifestyles that emphasize the quality of life and material sufficiency in a finite world.

8. *Advance the study of ecological sustainability and promote the open exchange and wide application of the knowledge acquired.*
   a. Support international scientific and technical cooperation on sustainability, with special attention to the needs of developing nations.
   b. Recognize and preserve the traditional knowledge and spiritual wisdom in all cultures that contribute to environmental protection and human well-being.
   c. Ensure that information of vital importance to human health and environmental protection, including genetic information, remains available in the public domain.

III. Social and Economic Justice

9. *Eradicate poverty as an ethical, social, and environmental imperative.*
   a. Guarantee the right to potable water, clean air, food security, uncontaminated soil, shelter, and safe sanitation, allocating the national and international resources required.
   b. Empower every human being with the education and resources to secure a sustainable livelihood, and provide social security and safety nets for those who are unable to support themselves.
   c. Recognize the ignored, protect the vulnerable, serve those who suffer, and enable them to develop their capacities and to pursue their aspirations.

10. *Ensure that economic activities and institutions at all levels promote human development in an equitable and sustainable manner.*
    a. Promote the equitable distribution of wealth within nations and among nations.
    b. Enhance the intellectual, financial, technical, and social resources of developing nations, and relieve them of onerous international debt.
    c. Ensure that all trade supports sustain-

able resource use, environmental protection, and progressive labor standards.

d. Require multinational corporations and international financial organizations to act transparently in the public goo[illegible] and hold them accountable for the co[illegible] sequences of their activities.

11. Affirm gender equality and equity [illegible] prerequisites to sustainable developm[illegible] and ensure universal access to educat[illegible] health care, and economic opportuni[illegible]

a. Secure the human rights of wom[illegible] girls and end all violence against t[illegible]

b. Promote the active participati[illegible] women in all aspects of econom[illegible] political, civil, social, and cultu[illegible] as full and equal partners, deci[illegible] makers, leaders, and beneficiar[illegible]

c. Strengthen families and ens[illegible] safety and loving nurture of a[illegible] members.

12. Uphold the right of all, wit[illegible] ination, to a natural and social en[illegible] supportive of human dignity, bodily health, and spiritual well-being, with special attention to the rights of indigenous peoples and minorities.

a. Eliminate discrimination in all its forms, such as that based on race, color, sex, sexual orientation, religion, language, and national, ethnic or social origin.

b. Affirm the right of indigenous peoples to their spirituality, knowledge, lands and resources and to their related practice of sustainable livelihoods.

c. Honor and support the young people of our communities, enabling them to fulfill their essential role in creating sustainable societies.

d. Protect and restore outstanding places of cultural and spiritual significance.

IV. Democracy, Nonviolence, and Peace

13. Strengthen democratic institutions at all levels, and provide transparency and accountability in governance, inclusive participation in decision making, and access to justice.

a. Uphold the right of everyone to receive clear and timely information on environmental matters and all develop[illegible] [illegible]tivities which are likely [illegible]y have an

[illegible] global civil [illegible]aningful [illegible]d individuals [illegible]n making.

[illegible]om of opinion, [illegible]bly, association,

[illegible]efficient access [illegible]dependent judi[illegible]g remedies and [illegible]al harm and the

[illegible] in all public and

[illegible]mmunities, [illegible]e for their environ[illegible]vironmental respon[illegible]s of government where they can be carried out most effectively.

14. Integrate into formal education and life-long learning the knowledge, values, and skills needed for a sustainable way of life.

a. Provide all, especially children and youth, with educational opportunities that empower them to contribute actively to sustainable development.

b. Promote the contribution of the arts and humanities as well as the sciences in sustainability education.

c. Enhance the role of the mass media in raising awareness of ecological and social challenges.

d. Recognize the importance of moral and spiritual education for sustainable living.

15. Treat all living beings with respect and consideration.

a. Prevent cruelty to animals kept in human societies and protect them from suffering.

b. Protect wild animals from methods of hunting, trapping, and fishing that cause extreme, prolonged, or avoidable suffering.

c. Avoid or eliminate to the full extent possible the taking or destruction of non-targeted species.

16. *Promote a culture of tolerance, nonviolence, and peace.*

a. Encourage and support mutual understanding, solidarity, and cooperation among all peoples and within and among nations.

b. Implement comprehensive strategies to prevent violent conflict and use collaborative problem solving to manage and resolve environmental conflicts and other disputes.

c. Demilitarize national security systems to the level of a non-provocative defense posture, and convert military resources to peaceful purposes, including ecological restoration.

d. Eliminate nuclear, biological, and toxic weapons and other weapons of mass destruction.

e. Ensure that the use of orbital and outer space supports environmental protection and peace.

f. Recognize that peace is the wholeness created by right relationships with oneself, other persons, other cultures, other life, Earth, and the larger whole of which all are a part.

The Way Forward

As never before in history, common destiny beckons us to seek a new beginning. Such renewal is the promise of these Earth Charter principles. To fulfill this promise, we must commit ourselves to adopt and promote the values and objectives of the Charter.

This requires a change of mind and heart. It requires a new sense of global interdependence and universal responsibility. We must imaginatively develop and apply the vision of a sustainable way of life locally, nationally, regionally, and globally. Our cultural diversity is a precious heritage and different cultures will find their own distinctive ways to realize the vision. We must deepen and expand the global dialogue that generated the Earth Charter, for we have much to learn from the ongoing collaborative search for truth and wisdom.

Life often involves tensions between important values. This can mean difficult choices. However, we must find ways to harmonize diversity with unity, the exercise of freedom with the common good, short-term objectives with long-term goals. Every individual, family, organization, and community has a vital role to play. The arts, sciences, religions, educational institutions, media, businesses, nongovernmental organizations, and governments are all called to offer creative leadership. The partnership of government, civil society, and business is essential for effective governance.

In order to build a sustainable global community, the nations of the world must renew their commitment to the United Nations, fulfill their obligations under existing international agreements, and support the implementation of Earth Charter principles with an international legally binding instrument on environment and development.

Let ours be a time remembered for the awakening of a new reverence for life, the firm resolve to achieve sustainability, the quickening of the struggle for justice and peace, and the joyful celebration of life.